# Standard Oil & Gas Abbreviator
### Sixth Edition

Compiled by
the Association of
Desk and Derrick Clubs

Copyright © 1973, 1986, 1994, 2001, 2006 by
PennWell Corporation
1421 South Sheridan/P.O. Box 1260
Tulsa, Oklahoma 74101

1-800-752-9764
sales@pennwell.com
www.pennwell.com
www.pennwellbooks.com

All rights reserved. No part of this book may be reproduced, stored in a retrieval system, or transcribed in any form or by any means, electronic or mechanical, including photocopying and recording, without the prior written permission of the publisher.

Printed in the United States of America

10  11  12  13  14      15  14  13  12  11

Managing Editor:  Marla M. Patterson
Book Designer:  Susan E. Ormston Thompson

# Contents

**Prefaces**
Sixth Edition . . . . . . . . . . . . . . . . . . . . . . . v
Fifth Edition . . . . . . . . . . . . . . . . . . . . . . . vii
Fourth Edition . . . . . . . . . . . . . . . . . . . . . . ix
Third Edition . . . . . . . . . . . . . . . . . . . . . . . xi
Second Edition . . . . . . . . . . . . . . . . . . . . . . xiii
First Edition . . . . . . . . . . . . . . . . . . . . . . . xv

**What is D&D?** . . . . . . . . . . . . . . . . . . . xvii

**Abbreviations with Definitions** . . . . . . . . . . . . . . 1

**Definitions with Abbreviations** . . . . . . . . . . . 149

**Abbreviations for Logging Tools and Services**
Baker Atlas Wireline Services . . . . . . . . . . . . 289
Schlumberger Well Services . . . . . . . . . . . . . 292
Hallibuton Log Service (HAL) . . . . . . . . . . . . 295
Log Heading Nomenclature . . . . . . . . . . . . . 297
Miscellaneous . . . . . . . . . . . . . . . . . . . . . 300

**Federal Environmental Acronyms** . . . . . . . . . . 308

**Pipe Coating Terminology and Definitions** . . . . . . 311

**Mnemonics**
Service Names . . . . . . . . . . . . . . . . . . . . 313
Computational Products . . . . . . . . . . . . . . . 318

**Abbreviations for Companies, Associations, and Organizations**
U. S. and Canada . . . . . . . . . . . . . . . . . . . 321
Outside U. S. and Canada . . . . . . . . . . . . . . 330

**Miscellaneous Information and Symbols**
Common Oilfield Spellings . . . . . . . . . . . . . . 357
API Standard Oil-Mapping Symbols . . . . . . . . . 365
Mathematical symbols and signs . . . . . . . . . . . 366

Standard Oil & Gas Abbreviator

Greek Alphabet . . . . . . . . . . . . . . . . . . . 367
Frequently Cited Chemical Abbreviations . . . . . . . . 368
Frequently Cited Aditive Abbreviations . . . . . . . . . 369
Frequently Cited Fluids Used as Cushion Abbreviations . 371
Directional Survey Calculation Methods . . . . . . . . 372
Directional Survey Processing Types . . . . . . . . . . 373
Directional Survey Type or Method to
 Determine Wellbore Path Deviation . . . . . . . . . 374
Lithology Abbreviations . . . . . . . . . . . . . . . . 375
System Equivalents . . . . . . . . . . . . . . . . . . 376
Metric-English System Conversion Factors . . . . . . . 377
Basic Conversion Factors . . . . . . . . . . . . . . . 379
Minerals Management Services Two Digit
 Area Prefix Standard . . . . . . . . . . . . . . . . 381
Petrophysical Curve Mneumonics . . . . . . . . . . . 383
**Universal Conversion Factors (CD-Rom Only)**
**Stratigraphic Nomenclature for Michigan (CD-Rom Only)**

# Preface to the Sixth Edition

As the energy and related industries continue to change in order to survive and continue to bring new technology to the forefront, so have the Association of Desk and Derrick Clubs. We are proud to be associated with the sixth edition of the D&D Standard Oil Abbreviator. It has become an indispensable tool within the oil and gas industry. Information included in this book has made writing tasks within energy and its related industries simpler, and has through the years added consistency to industry reports.

Since its inception the book has grown and now has several new sections. We are especially thankful to all of the energy and related industries that contributed to our sixth edition. There are five new sections included in this edition. They are as follows: Directional Survey Methods, Frequently Cited Additives, Frequently Cited Fluids, Lithology and Formation Names, MMS Two Digit Area Prefix Standards.

This book continues to be an ongoing tribute to the oil, gas, and energy industries that continue to develop and employ new technology. We are celebrating 55 years and hope to have fulfilled the need that our industry had in wanting an all-encompassing guide of terms used in the energy and allied industries.

Val Williams

# Preface to the Fifth Edition

The D&D Standard Oil Abbreviator is still a viable tool used by various individuals in the energy industries. New information has been added to this edition consistent to industry reports and writings.

New features for this edition include the full-color Michigan Stratigraphic Chart, and a fully electronic version of the D&D Abbreviator with the addition of Stephen Gerolde's Universal Conversion Factors on CD.

When the need for a fifth edition was announced, members of Desk and Derrick were asked to contribute any new changes, additions and deletions. As usual the members came through. I want to especially thank Elizabeth Dudley, Aramco Services Company; Dee Jansen, Stratland Exploration Company; Ellen Montgomery Coon, Great Lakes Directional Drilling; Mark Wollensack, Michigan Basin Geological Society and The Tulsa Geological Society for their input into this edition.

In this the 50th anniversary of the founding of Desk & Derrick we are proud to still be associated with such a publication as the D&D Standard Oil Abbreviator and proud to be a vital part of the Energy Industry.

Audrey M. Renegar

## Preface to the Fourth Edition

The D&D Standard Oil Abbreviator is an indispensable tool in the oil, gas, and energy industries. Information included in this book has made writing tasks within energy and its related industries simpler, and has through the years added consistency to industry reports.

The need for a fourth revision of this publication was announced in 1991. Immediately, all members of the Desk and Derrick Association were asked to contribute changes, additions, and/or deletions relating to their positions. Responses came not only from members, but also from industry businesses updating, correcting and adding according to each circumstance.

This fourth edition contains over 11,500 abbreviations and definitions used in energy-related industries.

Special thanks to Toni Stevens of Western Atlas International/Atlas Wireline Services (Anchorage, AK D&D) for her help in revising that portion of the Abbreviator; to Shirley Legaux (Lafayette, LA D&D) for the addition of the Federal Environmental Acronyms; Cindy Whitton (Tulsa, OK D&D), Ken Bell (Laurel, MS), Schlumberger Well Services, Marlene Meyers (Bakersfield, CA D&D), Linda Hill (New Orleans, LA D&D), and Linda Butler (N. Harris Montgomery County, TX D&D). A very special thank you to Bettye Hatcher Miller (Bellaire, TX D&D) for her continued interest and support of the D&D Abbreviator.

Company names have changed. Dresser-Atlas is now *Atlas Wireline Services*, and Welex is now *Halliburton Log Services* or *HAL*. Some entries may be repeated within company, miscellaneous and/or alphabetical listings; this is for the convenience of the user. All terms are alphabetized by abbreviation and by definition. Be sure to check the table of contents for the section most likely to contain the information which you work with. Once you begin to use the Abbreviator, you will find it an indispensable tool in your daily writing.

Additions to this edition include *Guide to Federal Environmental Acronyms*, *Computational Products Mnemonics*, *Service Name Mnemonics*, and *Log Heading Nomenclature*. Logging nomenclature standards have not been established; therefore, abbreviations for these tools and services are under a separate heading.

The fourth edition of the D&D Standard Oil Abbreviator is an ongoing tribute to the oil, gas, and energy industries that continues to develop and employ new technology and share it with the world. Consequently, there is a need served with each new edition of the Abbreviator—that of keeping our technical and everyday writing in step with the times. As the overall energy industry continues to change in order to survive, so have we of Desk and Derrick made this effort to simplify the daily work of those in the industry.

As a convenience for engineers, draftsmen, and others using this abbreviator, the back of the book contains standard map symbols through the courtesy of American Petroleum Institute, mathematical symbols, and the Greek alphabet.

The Association of Desk and Derrick Clubs hopes this book will continue to fill a need in the energy industry, not only to assist the secretary and the report writer, but to familiarize all newcomers with our unique way of doing things.

In this, the 45th anniversary year of the founding of Desk and Derrick, we are proud that we found a need and undertook to fill it—we hope we have successfully continued that good effort.

Jacqueline J. Brill

## Preface to the Third Edition

The *D&D Standard Oil Abbreviator* has become an indispensable tool within the oil and gas industry. The information it contains has made writing tasks much easier and has added uniformity.

When this revision was announced, the members of Desk and Derrick, as usual, came through with updated information from every aspect of the oil and gas industry. I have come to the conclusion that "we can abbreviate *everything*" if we put our minds to it. This Third Edition contains over 10,500 abbreviations and definitions for our industry.

Two new sections have been added. Within the Logging Section a new grouping was added for miscellaneous terms. Some are in the other groups, but this section also contains additional terms, thus alphabetizing all logging terms together. A new section was added—"Pipe Coating Terminology and Definitions"—and I am sure this will be a growing addition.

These abbreviations and terms have been mechanically alphabetized through the IBM 5520 System, which takes into account symbols, *i.e.*, hyphens, slashes, etc., and arranges the abbreviations and terms accordingly. Once you use the Abbreviator, you'll appreciate the ease of using this system.

I wish to thank all the members of D&D who made contributions to this new edition and also those people who helped arrange this information for publication.

My wish for this edition is to further carry on the hope of the first *D&D Standard Oil Abbreviator*—to fill a need.

Linda d'Allesandro Weatherly

# Preface to the Second Edition

The gratifying response to the first edition of the *D&D Oil Abbreviator*, which carried it through three printings, seemed to demand a second edition.

As in all undertakings, there are changes that "would be made if we had it to do over."

So the ladies of Desk and Derrick got behind the project a second time and submitted more data, pointed out earlier typographical errors, and suggested new features for this second edition.

This volume contains more abbreviations for words commonly used in the oil and gas industries than its predecessor. It also contains more abbreviations for company, association, agency, and society names. These are separated geographically, in a sense, with one list covering the U.S. and Canada and the second encompassing the rest of the world.

A major change is noted in the handling of logging tools and services. Because standards have not been established, and because abbreviations for these tools and services are rather complex, it was decided to contain them under a separate heading.

And as a convenience for secretaries, engineers, draftsmen, and others in the industry, the back of the book contains standard map symbols through the courtesy of American Petroleum Institute, mathematical symbols, and the Greek alphabet.

All in all, this second edition was designed to carry out the hope of the first—to fill a need.

## Preface to the First Edition

This book was developed from an idea proposed by Bettye Lynn Hatcher, Monahans, Texas, when she was president of the Monahans Desk and Derrick Club. She presented the idea in a feature story in the club's publication, "The Direct Line."

Her story described the plight of a young secretary on her first assignment with an oil company, trying to decipher the strange-looking language in a drilling report. With no "key" to unlock the mystery of this language, she was completely lost.

So here is the "key"—a listing of nearly every abbreviation used in the oil and gas industry, supplied by those most familiar with this peculiar brand of shorthand—the members of the Association of Desk and Derrick Clubs of North America.

Contributions came from these members in all geographical areas and in all branches of the industry from exploration to marketing. Every effort was made to make this first product as complete as possible, but omissions are bound to occur. Additional contributions were invited by *The Oil and Gas Journal*'s Book Department.

For convenience, the book is divided into sections. The first two of these are the largest and give abbreviations for words or phrases. The first section shows the abbreviation followed by its definition. In the second section this order is reversed.

The third section gives the American abbreviations for oil-related companies operating in countries other than the United States. This is followed by abbreviations for the many associations and societies related to the oil industry.

At the outset, the reader will note that there are some abbreviations that apply to more than one word or phrase. Out of context, this seems confusing, but when used in context, these abbreviations take on a clearer meaning and are easily understood by those familiar with the industry's everyday language.

On the matter of style, the contributions came in assorted forms; all capital letters, all lower-case letters, mixed caps, and lower-case, etc. This is not surprising because there are no standards, to follow yet, and each secretary is her own boss in this regard. But to introduce some consistency, the editors adopted a style based on simple logic: If an abbreviation is made up of the principal letters in a word, as in *bldg* for *building*, lower-case letters are used. If an abbreviation is made up of the first letters of several words (*WOC for waiting on cement*), capital letters are used. And to make the abbreviations as abbreviated as possible, periods are used only when their absence could lead to confusion.

It is hoped that this book fills a need in the industry, not only to help the secretarial side interpret the language, but to familiarize all newcomers to the industry with our peculiar ways of doing things.

# What is D&D?

**Purpose:** The purpose of the Association is to promote the education and professional development of individuals employed in or affiliated with the petroleum, energy and allied industries.

**Motto:** Greater Knowledge, Greater Service.

**Mission Statement:** To enhance and foster a positive image to the global community by promoting the contribution of the petroleum, energy, and allied industries through education by using all resources available.

**History:** A desire on the part of Inez Awty Schaeffer, Humble Oil and Refining Company, New Orleans, to meet other women in the Oil and Gas Industry, brought together in April, 1949, a group of women to form the first Desk and Derrick Club. The idea quickly spread to Jackson, Mississippi, Los Angeles, California, and Houston, Texas. The Association held its first annual convention in Houston, Texas in September, 1952. This history of The Association of Desk and Derrick Clubs is exceptional in that through the highs and lows of the petroleum industry, membership has fluctuated but remains strong and active.

Fifty-five years later, we are still going strong. With the introduction of male members in 1988, we, the men and women of ADDC, are continually striving to serve our purpose.

The purpose of The Association is sustained through monthly meetings of educational programs, as well as seminars and field trips to industry work sites. Industry study courses and publications including Fundamentals of Petroleum, Land and Leasing, and Petroleum Geology have been developed with the University of Texas Petroleum Extension (PETEX). CED(Continuing Education Units) are earned when courses and seminars are presented by PETEX-certified instructors.

**Contact:** ADDC for more information at 5153 East 51$^{st}$ St., Suite 107, Tulsa, Oklahoma 74135 or call the ADO Manager at 918-622-1749. E-Mail address is <u>adotulsa@swbell.net</u>. The ADDC has a comprehensive website at: <u>www.addc.org</u>.

# Abbreviations with Definitions

| | |
|---|---|
| A | abstract (i.e., A-10) |
| A | angstrom unit |
| A-Cem | acoustic cement |
| A&A | adjustments and allowances |
| A/ | acidized with |
| A/C | air conditioning |
| A/CLR | air cooler |
| A/P | accounts payable |
| A/R | accounts receivable |
| AA | after acidizing, as above |
| AAR | Association of American Railroads |
| ABC | Audit Bureau of Circulation |
| abd | abandoned |
| abd loc | abandoned location |
| abd-gw | abandoned gas well |
| abd-ow | abandoned oil well |
| abd-ogw | abandoned oil & gas well |
| ABHL | absolute bottom-hole location |
| ABM | Atlas Bradford modified |
| abrsi jet | abrasive jet |
| abs | absolute |
| ABS | acrylonitrile butadiene styrene rubber |
| absm | absorption |
| abst | abstract |
| abt | about |
| abun | abundant |
| abv | above |
| ac | acid |
| ac | acre(s), acreage |
| AC | alternating current |

1

| | |
|---|---|
| AC | Austin chalk |
| ac-ft | acre-feet |
| ACC | access |
| ACCEL | accelerometers |
| ACCESS | accessory, accessories |
| acct | account(ing) |
| accum | accumulative, accumulator |
| acd | acidize (ed) (ing) |
| acfr | acid fracture treatment |
| A/CLD | air cooled |
| ACLR | air cooler |
| ACM | acid-cut mud |
| acrg | acreage |
| ACS | American Chemical Society |
| ACSR | aluminum conductor steel reinforced |
| ACT | actual |
| ACT | actuated, actuator |
| ACT | automatic custody transfer |
| ACW | acid-cut water |
| AD | actual drilling |
| AD | authorized depth |
| ADC | actual drilling cost |
| add | additive |
| addl | additional |
| ADH | adhesive |
| adj | adjustable |
| adm | administration, administrative |
| ADOM | adomite |
| ADP | automatic data processing |
| adpt | adapter |
| adspn | adsorption |
| ADT | actual drilling time |
| advan | advanced |
| AER | aeration, aerator |

# Abbreviations with Definitions

| | |
|---|---|
| AF | acid frac |
| AF | after fracture |
| AF/CLR | after cooler |
| AF/COND | after condenser |
| AFC | Authorization for Commitment |
| AFC | Authorized for Construction |
| AFD | auxiliary flow diagram |
| AFE | Authorization for Expenditure |
| affd | affirmed |
| afft | affidavit |
| AFIT | after federal income tax |
| AFP | average flowing pressure |
| AFRA | average freight rate assessment |
| AG | agitator |
| AGGR | aggregate |
| agim | agglomerate |
| AIR | average injection rate |
| AIR COND | air conditioning |
| AJT | actual jetting time |
| AL | aluminum |
| AL | artificial lift |
| Alb | Albany |
| alc | alcoholic |
| ALCOA | Aluminum Company of America |
| alg | algae |
| alg | along |
| ALIGN | alignment (ing) |
| alk | alkaline, alkalinity |
| alkyl | alkylate, alkylation |
| ALLOW | allowable, allowance |
| alm | alarm |
| ALOC | allocation |
| alt | alternate |
| ALT | altitude |

| | |
|---|---|
| ALY | alloy |
| amb | ambient |
| AMI | area of mutual interest |
| AMM | ammeter |
| amor | amorphous |
| amort | amortization |
| AMP | American melting point |
| amp | ampere |
| amp hr | ampere hour |
| amph | amphipore |
| Amph | Amphistegina |
| AMR | addition or modification reque |
| amt | amount |
| an | annulus |
| anal | analysis, analytical |
| ANC | Anchor (age) |
| ang | angle, angular |
| Angul | Angulogerina |
| anhy | anhydrite, anhydritic |
| anhyd | anhydrous |
| ANR | amount not reported |
| ANS | Alaskan North Slope |
| ANUB | annubar |
| ANUC | annunciator |
| ANYA | allowable not yet available |
| AOF | absolute open flow potential (gas well) |
| app | appears, appearance |
| APPAR | apparatus |
| appd | approved |
| appl | appliance |
| appl | applied |
| applic | application |
| approx | approximate (ly) |
| apr | apparent (ly) |

## Abbreviations with Definitions

| | |
|---|---|
| APR | average penetration rate |
| apt | apartment |
| aq | aqueous |
| AQCR | air quality control region |
| AQMA | air quality maintenance area |
| AR | acid residue |
| Ara | Arapahoe |
| arag | aragonite |
| Arb | Arbuckle |
| arch | architectural |
| Archeo | Archeozoic |
| aren | arenaceous |
| arg | argillaceous |
| arg | argillite |
| ark | arkose(ic) |
| Arka | Arkadelphia |
| arm | armature |
| arnd | around |
| ARO | after receipt of order (purchasing term) |
| ARO | at rate of |
| arom | aromatics |
| ARR | arrange (ed) (ing) (ment) |
| AS | after shot |
| AS | anhydrite stringer |
| AS&W ga | American Steel & Wire gauge |
| ASA | American Standards Association |
| ASAP | as soon as possible |
| asb | asbestos |
| asbr | absorber |
| ASD | abandon-salvage deferred |
| asgmt | assignment |
| Ash | ashern |
| ASO | acid-soluble oil |
| asph | asphalt. asphaltic |

| | |
|---|---|
| assgd | assigned |
| assn | association |
| assoc | associate (d) (s) |
| asst | assistant |
| assy | assembly |
| astn | asphaltic stain |
| ASW | adjustable spring wedge |
| AT | acid treat (ment) |
| AT | after treatment |
| AT | all thread |
| ATT | after the tanks |
| At | Atoka |
| at | atomic |
| at wt | atomic weight |
| ATC | after top center |
| ATD | approved total depth |
| ATF | automatic transmission fluid |
| atm | atmosphere, atmospheric |
| ATP | Authorization to Proceed |
| ATP | average treating pressure |
| ATP | average tubing pressure |
| ATT | attach (ed) (ing) (ment) |
| att | attempt(ed) |
| atty | attorney |
| aud | auditorium |
| Aus | Austin |
| auth | authorized |
| auto | automatic |
| auto | automotive |
| autogas | automotive gasoline |
| aux | auxiliary |
| AV | annular velocity |
| AV | Aux Vases sand |
| av | aviation |

## Abbreviations with Definitions

| | |
|---|---|
| avail | available |
| AVC | automatic volume control |
| avg | average |
| avgas | aviation gasoline |
| AW | acid water |
| AWD | award |
| AWG | American Wire Gauge |
| awtg | awaiting |
| az | azimuth |
| aztrop | Azeotrophic |

## B

| | |
|---|---|
| B | billion |
| B | bulletin |
| B&B | bell and bell |
| B &CB | beaded and center beaded |
| B&S | bell and spigot |
| B Hn | Big Horn |
| B slt | base of the salt |
| B. In. | Big Injun |
| B. Ls | Big Lime |
| B. Riv | Black River |
| B. Bl | Base Blane |
| B.E. | beveled end |
| BS&W | basic sediment and water |
| B&B | bent & bowed pipe |
| B&F | bell and flange |
| B&S | Brown & Sharpe (gauge) |
| B/ | base |
| B/ | bottom of given formation (i.e., B/Frio) |
| B/B | back to back |
| B/B | barrels per barrel |

| | |
|---|---|
| B/dry | bailed dry |
| B/hr | barrels per hour |
| B/JT | ball joint |
| B/L | bill of lading |
| B/M | bill of material |
| b/off | buck-off |
| b/on | buck-on |
| B/S | back scuttled |
| B/S | base salt |
| B/S | bending schedule |
| B/S | bill of sale |
| B/SD | barrels per stream (refinery) |
| B/VESS | bulk vessel |
| B/Vlv | ball valve |
| BA | barrels of acid |
| bail | bail (ed) |
| BAL | balance |
| Ball | Balltown sand |
| bar | barite (ic) |
| Bar | Barlow Lime |
| bar | barometer, barometric |
| BAR | barrels acid residue |
| Bark Crk | Barker Creek |
| Bart | Bartlesville |
| base | basement (granite) |
| bat | battery |
| BAT | before acid treatment |
| Bate | Bateman |
| BAW | barrels acid water |
| BAWPD | barrels acid water per day |
| BAWPH | barrels acid water per hour |
| BAWUL | barrels acid water under load |
| BB | bridged back |
| BB fraction | butane-butene fraction |

# Abbreviations with Definitions

| | |
|---|---|
| BBE | bevel both ends |
| bbl | barrel |
| BC | barrels of condensate |
| Bcf | billion cubic feet |
| Bcfd | billion cubic feet per day |
| BCPD | barrels condensate per day |
| BCPH | barrels condensate per hour |
| BCPMM | barrels condensate per million |
| BD | barrels of distillate |
| BD | blowdown |
| bd | board |
| BD | breakdown |
| BD | budgeted depth |
| bd ft | board foot (feet) |
| BD-MLW | barge deck to mean low water |
| Bd'A | Bois d' Arc |
| BDA | breakdown acid |
| BDF | broke (break) down formation |
| BDL | bundle |
| BDNG | bedding |
| BDO | barrels diesel oil |
| BDP | breakdown pressure |
| BDPD | barrels distillate per day |
| BDPH | barrels distillate per hour |
| BDT | blow-down test |
| Be | Baumé |
| Be | Berea |
| be | box end |
| Bear Riv | Bear River |
| bec | becoming |
| Beck | Beckwith |
| Bel | Beldon |
| Bel C | Belle City |
| Bel F | Belle Fourche |

| | |
|---|---|
| Belm | belemnites |
| Ben | Benoist (Bethel) sand |
| Ben | Benton |
| Bent | bentonite |
| berm | berm, sloped wall to keep out flooding |
| bev | bevel (ed) |
| BF | barrels fluid |
| BF | blind flange |
| bf | buff |
| BFIT | before federal income tax |
| BFL | baffle |
| BFO | barrels frac oil |
| BFPD | barrels fluid per day |
| BFPH | barrels fluid per hour |
| BFW | bailer feed water |
| BFW | barrels formation water |
| BFW | barrels fresh water |
| BFW | boiler feed water |
| BH | bottom hole |
| BHA | bottom-hole assembly |
| BHC | bottom-hole choke |
| BHCS | borehole compensated sonic |
| BHF | Bradenhead Flange |
| BHFP | bottom-hole flowing pressure |
| BHL | bottom-hole location |
| BHM | bottom-hole money |
| BHN | Brinell hardness number |
| BHO | bottom-hole orientation |
| BHP | bottom-hole pressure |
| bhp | brake horsepower |
| bhp-hr | brake horsepower-hour |
| BHPC | bottom-hole pressure closet (*see also* SIBHP *and* BHSIP) |
| BHPF | bottom-hole pressure flowing |

## Abbreviations with Definitions

| | |
|---|---|
| BHPS | bottom-hole pressure survey |
| BHSIP | bottom-hole shutin pressure |
| BHT | bottom-hole temperature |
| BID SUM | bid summary |
| Big. | Bigenerina |
| Big. f. | Bigenerina floridana |
| Big. h. | Bigenerina humblei |
| Big. nod. | Bigenerina nodosaria |
| BIN | binary |
| bio | biotite |
| bit | bitumen, bituminous |
| BKFLSH | back flush |
| bkr | breaker |
| BKWSH | backwash |
| BL | barrels load |
| bl | blue |
| BL&AW | barrels load & acid water |
| Bl/Cb | blast cabinet |
| BL/JT | blast joint |
| BLC | barrels load condensate |
| BLCPD | barrels load condensate per day |
| BLCPH | barrels load condensate per hour |
| bld | bailed |
| bld | bleeding |
| bld gas | bleeding gas |
| BLD FLG, BF | blind flange |
| bldg | building |
| bldg drk | building derrick |
| bldg rds | building roads |
| bldo | bleeding oil |
| bldrs | boulders |
| BLE | bevel large end |
| blg | bailing |
| Blin | Blinebry |

| | |
|---|---|
| blk | black |
| blk | block |
| Blk Lf | Black Leaf |
| Blk Li | Black Lime |
| blk lnr | blank liner |
| BLND | blend (ed) (er) (ing) |
| BLO | barrels load oil |
| blo | blow |
| BLOPD | barrels load oil per day |
| BLOPH | barrels load oil per hour |
| BLOR | barrels load oil recovered |
| Blos | blossom |
| BLOTBR | barrels load oil to be recovered |
| BLOYTR | barrels load oil yet to recover |
| BLPD | barrels of liquid per day |
| blr | bailer |
| BLR | boiler |
| blts | bullets |
| BLW | barrels load water |
| BLWPD | barrels load water per day |
| BLWPH | barrels load water per hour |
| BLWR | blower |
| BLWTR | barrels load water to recover |
| BM | barrels mud |
| BM | benchmark |
| BM | Black Magic (mud) |
| BMEP | brake mean effective pressure |
| BMI | black malleable iron |
| bmpr | bumper |
| bn | brown |
| bnd | band (ed) |
| bndry | boundary |
| BNO | barrels new oil |
| BNW | barrels new water |

| | |
|---|---|
| bnz | benzene |
| BO | backed out (off) |
| BO | barrels oil |
| BO | blew out |
| BO | blocked off |
| BO | free-point back off |
| BOCD | barrels oil per calendar day |
| BOCS | Basal Oil Creek Sand |
| BOD | biochemical oxygen demand |
| Bod | Bodcaw |
| BOE | bevel one end |
| BOE | blowout equipment |
| Bol. | Bolivarensis |
| Bol. a. | Bolivina a. |
| Bol. flor. | Bolivina floridana |
| Bol. p. | Bolivina perca |
| Bonne | Bonneterre |
| BOP | blowout preventer |
| BOPD | barrels oil per day |
| BOPE | blowout preventer equipment |
| BOPH | barrels oil per hour |
| BOPPD | barrels oil per producing day |
| BOS | brown oil stain |
| bot | bottom |
| BP | back pressure |
| BP | base Pennsylvania |
| BP | Bearpaw |
| BP | boiling point |
| BP | bridge plug |
| BP | bulk plant |
| BP | bull plug |
| BP Mix | butane and propane mix |
| BP/CLR | bypass cooler |
| BPD | barrels per day |

## Standard Oil & Gas Abbreviator

| | |
|---|---|
| BPH | barrels per hour |
| BPLO | barrels of pipeline oil |
| BPLOPD | barrels of pipeline oil per day |
| BPM | barrels per minute |
| BPSD | barrels per stream day |
| BPV | back-pressure valve |
| BPWPD | barrels per well per day |
| BR | building rig |
| BR | building road |
| brach | brachiopod |
| BRC | brace (ing) (ed) |
| brec | breccia |
| BRFL/V | butterfly valve |
| brg | bearing |
| Brid | bridger |
| brit | brittle |
| brk | break (broke) |
| brkn | broken |
| brkn sd | broken sand |
| BRKR | breaker |
| BRKS | brakes |
| brksh | brackish (water) |
| brkt(s) | brackets(s) |
| Brn Li | brown lime |
| brnsh | brownish |
| brn or br | brown |
| brn sh | brown shale |
| Brom | bromide |
| brtl | brittle |
| bry | bryozoa |
| BS | ball sealers |
| BS | basic sediment |
| BS | Bone Spring |
| BS | bottom sediment |

## Abbreviations with Definitions

| | |
|---|---|
| BS | bottom settlings |
| BS&W | bottom (basic) sediment & water |
| Bscf | billion standard cubic feet |
| Bscf/d | billion standard cubic feet per day |
| BSE | bevel small end |
| BSFC | brake specific fuel consumption |
| BSHG | bushing |
| BSI | British Standards Institution |
| bskt | basket |
| bsl | basal |
| bsmt | basement |
| BSPL | base plate |
| BSTR | booster |
| BSUW | black sulfur water |
| BSW | barrels salt water |
| BSWPD | barrels salt water per day |
| BSWPH | barrels salt water per hour |
| BT | Benoist (Bethel) sand |
| BTC | buttress thread coupling |
| BTDC | before top dead center |
| BTFL/V | butterfly valve |
| btm (d) | bottom (ed) |
| btm chk | bottom choke |
| btry | battery |
| BTU | British thermal unit |
| btw | between |
| BTWLD | butt weld |
| BTX | benzene toluenexylene (unit) |
| bu | bushel |
| Buck | Buckner |
| Buckr | buckrange |
| Bul. text. | Buliminella textularia |
| Bull W | Bullwaggon |
| Bum. | bottom-hole pressure bomb |

| | |
|---|---|
| bunr | burner |
| Burg | Burgess |
| butt | buttress thread |
| BUZ | buzzer |
| BV | block valve |
| BV/WLD | beveled for welding |
| BW | barrels of water |
| BW | boiled water |
| BW | butt weld |
| BW ga | Birmingham (or Stubbs) iron wire gauge |
| BW/D | barrels of water per day |
| Bwg | Birmingham wire gauge |
| BWL | barrel water load |
| BWL | body wall loss |
| BWOL | barrels water over load |
| BWPD | barrels water per day |
| BWPH | barrels water per hour |
| bx | box (es) |
| BYP | bypass |

## C

| | |
|---|---|
| C | Celsius |
| C | center (land description) |
| C | centigrade |
| c | coarse (ly) |
| C | core hole |
| C&F | cost and freight |
| C&P | cellar & pits |
| C&W | coat and wrap (pipe) |
| C to C | center to center |
| C to E | center to end |
| C to F | center to face |

## Abbreviations with Definitions

| | |
|---|---|
| C.I.F. | cost insurance and freight |
| C.O.P. | completed on pump |
| C&A | compression and absorption plant |
| C&C | circulate & condition |
| C&CH | circulated and conditioned hole |
| C&CM | circulated and conditioned mud |
| C&R | circulate and reciprocate |
| C/ | contractor (i.e., C/John Doe) |
| C/A | commission agent |
| C/BM | crawl beam |
| C/H | cased hole |
| C/L | center line |
| c/o | care of |
| C/W | complete with |
| CA | corrosion allowance |
| CAB | cabinet |
| Cadd | Caddell |
| CAG | cut across grain |
| cal | calcite, calcitic |
| cal | caliche |
| CAL | caliper log |
| cal | caliper survey |
| cal | calorie |
| Calc | calcareous, calcerenite |
| Calc | calcium |
| Calc | calculate (ed), calculation |
| calc gr | calcium-base grease |
| cale | calceneous |
| CALIBR | calibrate, calibration |
| Camb | cambrian |
| Calv | Calvin |
| CaO | calcium oxide |
| Cane Riv | Cane River |
| Cany | canyon |

| | |
|---|---|
| CaO | calcium oxcide |
| CAOF | calculated absolute open flow |
| cap | capacitor |
| cap | capacity |
| Cap | Captain |
| Car | Carlile |
| carb | carbonaceous |
| carb test | carbontetrachloride |
| Carm | Carmel |
| Casp | Casper |
| CAT | carburetor air temperature |
| Cat | Catahoula |
| CAT | catalog |
| CAT | catalyst, catalytic |
| cat ckr | catalytic cracker |
| Cat Crk | Cat Creek |
| cath | cathodic |
| caus | caustic |
| cav | cavity |
| CB | changed (ing) bits |
| CB | continuous blowdown |
| CB | core barrel |
| CB | counterbalance (pumping equip.) |
| CBL | cable (ing) |
| CBM | Coal bed methane |
| CBU | circulate bottoms up |
| CC | calcium chloride |
| CC | carbon copy |
| CC | casing cemented (depth) |
| CC | closed cup |
| cc | cubic centimeter |
| C-Cal | contact caliper |
| ccBU | circulate bottoms up |
| CCHF | center of casinghead flange |

## Abbreviations with Definitions

| | |
|---|---|
| Cck | casing choke |
| CCL | casing collar locator |
| CCLGO | cat-cracked light gas oil |
| CCM | condensate-cut mud |
| CCP | central compressor plant |
| CCP | critical compression pressure |
| CCPR | casing collar perforating record |
| CCR | Conradson carbon residue |
| CCR | critical compression ratio |
| CCS | California Coordinate System |
| CCS | cast carbon steel |
| CCS | computer control system |
| CCU | catalytic cracking unit |
| ccw | counterclockwise |
| CD | calendar day |
| CD | cold drawn |
| CD | contract depth |
| CD PL | cadmium plate |
| CDB | cement dump bailer |
| CDB | common data base |
| CDBTF | common data base task force |
| CDL | cut drilling line |
| CDM | continuous dipmeter survey |
| CDO | certified drawing outline |
| CDP | Central Delivery Point |
| Cdr Mtn | Cedar Mountain |
| CDS | continuous directional service |
| cdsr | condenser |
| Cdy | Cody (Wyoming) |
| cell | cellar |
| cell | cellular |
| CEM | Cement (ed) |
| CEMF | counter electromotive force |
| Ceno | Cenozoic |

Standard Oil & Gas Abbreviator

| | |
|---|---|
| cent | centralizers |
| centr | centrifugal |
| ceph | cephalopod |
| Cert. ex. | Ceratobulimina eximia |
| CERT | certified |
| CET | cement evaluation |
| CF | casing flange |
| CF | clay filled |
| Cf | Cockfield |
| CF | cold finished |
| CFB&G | companion flange bolt and gasket |
| CFBO | companion flanges bolted on |
| cfd | cubic feet per day |
| CFE | contractor furnished equipment |
| cfg | cubic feet gas |
| cfgd | cubic feet gas per day |
| cfgh | cubic feet gas per hour |
| CFM | Continuous flowmeter |
| cu ft | cubic foot |
| cu ft/bbl | cubic feet per barrel |
| cu ft/min | cubic feet per minute |
| cu ft/sec | cubic feet per second |
| cfp | cubic feet per pound |
| CFR | cement friction reducer |
| CFR | cement friction retarder |
| CFRC | Coordinating Fuel Research Committee |
| cfs | cubic feet per second |
| CG | center of gravity |
| cg | centigram |
| cg | coarse grained |
| cg | coring |
| CG | corrected gravity |
| cglt | conglomerate, conglomeritic |
| C-gr | coarse grained |

| | |
|---|---|
| cgs | centimeter-gram-second-system |
| CH | casinghead (gas) |
| ch | chert |
| ch | choke |
| CH | closed hole |
| CH | core hole |
| CH OP | chain operated |
| chal | chalcedony |
| CHAM | chamfer |
| Chapp | Chappel |
| CHAR | characteristics |
| Char | Charles |
| Chatt | Chattanooga shale |
| CHD | closed hydrocarbon drain |
| chem. | chemical, chemist, chemistry |
| chem. prod | chemical products |
| Cher | Cherokee |
| Ches | Chester |
| CHF | casinghead flange |
| CHG | casinghead gas |
| chng | change (ed) (ing) |
| Chngd DP | changed drillpipe |
| chrg | charge (ed) (ing) |
| Chim H | Chimney Hill |
| Chim R | Chimney Rock |
| Chin | Chinle |
| chit | chitin (ous) |
| chk | chalk |
| chk | choke |
| Chkbd | checkerboard |
| chkd | checked |
| CHKD PL | checkered plate |
| CHKV | check valve |
| chky | chalky |

| | |
|---|---|
| chl | chloride (s) |
| chl | chloritic |
| chl log | chlorine long |
| CHLR | chlorinator |
| CHMBR | chamber |
| CHNL | channel |
| Chou | Chouteau lime |
| CHP | casinghead pressure |
| chrm | chairman |
| chromat | chromatograph |
| chrome | chromium |
| cht | chart |
| cht | chert |
| chty | cherty |
| Chug | chugwater |
| CI | cast iron |
| CI | contour interval (map) |
| CI engine | compression-ignition engine |
| Cib. | Cibicides |
| Cib. h. | Cibicides hazzardi |
| CIBP | cast-iron bridge plug |
| CIE | crude industrial ethanol |
| Cima | Cimarron |
| CIP | cement in place |
| CIP | closed-in pressure |
| cir | circle |
| cir | circuit |
| cir | circular |
| cir mils | circular mills |
| circ | circulate, circulating, circulation |
| Cis | Cisco |
| ck | cake |
| ck | check |
| Ck Mtn | Cook Mountain |

| | |
|---|---|
| cksn | chicksan |
| CL | carload |
| cl | centiliter |
| CL | class |
| Clag | Clagget |
| Claib | Claiborne |
| Clarks | Clarksville |
| clas | clastic |
| CLASS | classification |
| Clav | Clavalinoides |
| Clay | Clayton |
| Clay | Claytonville |
| Cleve | Cleveland |
| Clfk | Clearfork |
| CLFR | clarifier |
| CLG | cooling |
| CLG/TWR | cooling tower |
| Cliff H | Cliff House |
| CLKG | caulking |
| CLMP | canvas-lined metal petal basket |
| cln (d) (g) | clean (ed) (ing) |
| Clov | clovery |
| clr | clear, clearance |
| CLR | cooler |
| CLR/TWR | cooling tower |
| clrg | clearing |
| clsd | closed |
| CLTR | collector |
| clyst | Claystone |
| $Cl_2$ | chlorine |
| cm | centimeter |
| cm/sec | centimeters per second |
| CMA | acoustic caliper |
| CMC | sodium carboxymethylcellulose |

| | |
|---|---|
| Cmchn | comanchean |
| CMPARTR | comparator |
| CMPD | compound |
| Cmpt | compact |
| cmt(d)(g)(r) | cement (ed) (ing) (er) |
| CN | cetane number |
| CN/BD | control building |
| cncn | concentric |
| CND | conduit |
| CNL | compensated neutron log |
| CNR | corner |
| cntf | centrifuge |
| cntl | control (s) |
| CNTN | containment |
| cntr | center (ed) |
| cntr | container |
| cntr | controller |
| CNTWT | counter weight |
| Cnty | county |
| cnvr | conveyor |
| CO | carbon monoxide |
| CO | carbon oxygen |
| CO | circulated out |
| CO | clean out |
| CO | cleaning out, cleaned out |
| Co | company |
| CO | crude oil |
| CO&S | clean out & shoot |
| Co. Op. | company operated |
| Co. Op. S.S. | company-operated service stations |
| co-op | cooperative |
| COBOL | Common Business-Oriented Language |
| COC | Cleveland open cup |
| Coco | Coconino |

## Abbreviations with Definitions

| | |
|---|---|
| COD | chemical oxygen demand |
| Cod | Codell |
| coef | coefficient |
| COF | calculated open flow (potential) |
| COG | coke oven gas |
| COL | collar |
| COL | colored |
| COL | column |
| Col ASTM | Color American Standard Test Method |
| Cole Jct | Coleman Junction |
| coll | collect (ed) (ing) (ion) |
| colr | collar |
| Com | Comanche |
| Com | Comatula |
| com | common |
| Com Pk | Comanche Peak |
| comb | combined, combination |
| COMB | combustion |
| coml | commercial |
| comm | commenced |
| comm | commission |
| comm | communication |
| comm | community |
| commr | commissioner |
| comp | complete (ed) (tion) |
| comp nat | completed natural |
| compnts | components |
| compr | compressor |
| compr st | compressor station |
| compt | compartment |
| COMPT | components |
| COMPTR | computer |
| COMT | comment |
| COMUT | commutator |

| | |
|---|---|
| con | consolidated |
| conc | concentrate |
| conc | concentric |
| conc | concrete |
| conch | conchoidal |
| concl | conclusion |
| cond | condensate |
| cond | condition (ed) (ing) |
| condr | conductor (pipe) |
| condt | conductivity |
| conf | confidential |
| conf | confirm (ed) (ing) |
| confl | conflict |
| cong | conglomerate (itic) |
| conn | connection |
| cono | conodonts |
| consol | consolidated |
| const | constant |
| const | construction |
| consv | conserve, conservation |
| cont (d) | continue (ed) |
| contam | contaminated, contamination |
| contr | contractor |
| contr resp | contractor responsibility |
| contrib | contribution |
| conv | converse |
| CONVT | convector, convection |
| COOH | coming out of hole |
| coord | coordinate |
| COP | crude oil purchasing |
| coq | coquina |
| cor | corner |
| Corp | corporation |
| corr | correct (ed) (ion) |

## Abbreviations with Definitions

| | |
|---|---|
| corr | corrosion |
| corr | corrugated |
| correl | correlation |
| corres | correspondence |
| COSH | hyperbolic cosine |
| COTD | cleaned out to total depth |
| COTH | hyperbolic cotangent |
| Cott G | Cottage Grove |
| Counc G | Council Grove |
| $CO_2$ | carbon dioxide |
| CP | casing point |
| CP | casing pressure |
| cp | centipose (s) |
| cp | chemically pure |
| CP | correlation point |
| Cp Colo | Camp Colorado |
| CPA | certified public accountant |
| CPC | casing pressure closed |
| cp'd | cemented through perforations |
| CPF | casing pressure flowing |
| CPF | central processing facility |
| CPG | cost per gallon |
| cplg | coupling |
| CPM | cycles per minute |
| CPO | confirming telephone order (purchasing term) |
| CPR | Cooper River Meridian (Alaska) |
| CPS | cycles per second |
| CPSI | casing pressure shut in |
| CR | cold rolled |
| CR | compression ratio |
| CR | Cow Run |
| CR Con | carbon residue (Conradson) |
| cr moly | chrome molybdenum |
| cr (d) (g) | core (ed) (ing) |

| | |
|---|---|
| CRA | cased reservoir analysis |
| CRA | chemically retarded acid |
| crbd | crossbedded |
| CRCMF | circumference |
| crd | cored |
| CRDL | cradle (s) |
| cren | crenulated |
| Cret | Cretaceous |
| crg | coring |
| Crin | crinoid (al) |
| Cris | Christellaria |
| crit | critical |
| crk | creek |
| crkg | cracking |
| Crkr | cracker |
| CRN | crane |
| crn blk | crown block |
| crnk | crinkled |
| Crom | Cromwell |
| crs | coarse (ly) |
| CRS | cold-rolled steel |
| CRS | cross |
| crs-xln | coarse crystalline |
| CRT | cathode ray tube |
| CRV | curve |
| crypto-xln | cryptocrystalline |
| cryst | crystalline |
| CS | carbon steel |
| CS | casing seat |
| CS | cast steel |
| cs | centistokes |
| CSA | casing set at |
| CSCH | hyperbolic constant |
| cse gr | coarse grained |

# Abbreviations with Definitions

| | |
|---|---|
| csg | casing |
| CSK | countersink |
| CSL | center section line |
| CSL | county school lands |
| $CS_2$ | carbon disulfide |
| CT | cable tools |
| CT | cooling tower |
| CTC | consumer tank car |
| ctd | coated |
| CTD | corrected total depth |
| ctg(s) | cuttings |
| CTHF | center of tubing flange |
| Ctlmn | Cattleman |
| CTM | cable tool measurement |
| ctn | carton |
| Ctnwd | Cottonwood |
| CTO | confirming telephone order (purchasing term) |
| CTP | cleaning to pits |
| ctr | center |
| CTS | cement to surface |
| CTT | consumer transport truck |
| CTU | coiled tubing unit |
| CTW | consumer tank wagon |
| CU | clean up |
| cu | cubic |
| cu cm | cubic centimeter |
| cu in. | cubic inch |
| cu m | cubic meter |
| cu yd | cubic yard |
| CUB | cubical |
| culv | culvert |
| cum | cumulative |
| Cur | Curtis |
| cush | cushion |

| | |
|---|---|
| CUST | customer |
| Cut B | cut bank |
| Cut Oil | cutting oil |
| Cut Oil Act Sul dk | cutting oil active-sulfurized-dark |
| Cut Oil Act Sul trpt | cutting oil active-sulfurized-transparent |
| Cut Oil Inact S | cutting oil inactive-sulfurize |
| Cut Oil Sol | cutting oil soluble |
| Cut Oil St Mrl | cutting oil straight mineral |
| cutbk | cutback |
| Cutl | Cutler |
| CV | control valve |
| CV | Cotton Valley |
| cvg(s) | caving (s) |
| CVR | cover |
| CVTR | convert (er) (ed) |
| cw | clockwise |
| CW | continuous weld |
| CW | cooling water |
| CWE | cold water equivalent |
| CWP | cold working pressure |
| CWR | cooling water return |
| CWS | cooling water supplying |
| cwt | hundredweight |
| CX | crossover |
| Cy Sd | Cypress Sand |
| Cyc | cyclamina |
| CYC | cyclone |
| Cycl | cyclone |
| Cycl canc. | Cyclamina cancellata |
| cyl | cylinder |
| cyn | canyon |
| Cyp | Cypridopsis |
| Cz | Carrizo |

## D

| | |
|---|---|
| D | day |
| D | Deca |
| D | development |
| D | dual |
| D&A | dry and abandoned |
| D&B | Dun & Bradstreet |
| D&C | drill and complete |
| D.I. | diesel index |
| D.O. | division order |
| d-d-1-s-1-e | dressed dimension one side and one edge |
| d-d-4-s | dressed dimension four side |
| d-1-s | dressed one side |
| D-2 | Diesel No. 2 |
| d-2-s | dressed two sides |
| d-4-s | dressed four sides |
| d/b/a | doing business as |
| D/D | day to day |
| D/L | density log |
| D/O | division office |
| D/P | differential pressure |
| D/P | drill (ed) (ing) plug |
| D/S | data sheet |
| D/T | driller's top |
| D/T | drilling tender |
| DA | daily allowable |
| DA | Dresser Atlas |
| DA | drift angle |
| DAIB | daily average injection barrels |
| Dak | Dakota |
| Dan | Dantzler |

| | |
|---|---|
| Dar | Darwin |
| DAR | discovery allowable requested darcy (darcies not abbreviated) |
| dat | datum |
| db | decibel |
| DB | diamond bit |
| DB | drilling break |
| DBA | depth bracket allowable |
| DBL | double |
| DBO | dark brown oil |
| DBOS | dark brown oil stains |
| DC | delayed coker |
| DC | development well, carbon dioxide |
| DC | diamond core |
| DC | digging cellar or digging cellar and slush pits |
| DC | direct current |
| DC | drill collar |
| DC | dually completed |
| DCB | diamond core bit |
| DCLSP | digging slush pits |
| DCM | distillate-cut mud |
| DCS | distributed control system |
| DCTR | detector |
| dd | dead |
| DD | degree day |
| DD | deviation degrees |
| DD | drilling deeper |
| DD | dyna-drilling |
| DDD | dry desiccant dehydrator |
| DDT | dichloro diphenyltrichloroethane |
| DE | double end |
| DEA | diethanolamine |
| DEA unit | diethanolamine unit |
| Deadw | Deadwood |

## Abbreviations with Definitions

| | |
|---|---|
| deaer | deaerator |
| deasph | deasphalting |
| debutzr | debutanizer |
| dec | decimal |
| decl | decline |
| decr | decrease (ed) (ing) |
| deethzr | deethanizer |
| defl | deflection |
| Deg | Degonia |
| deg | degree (s) |
| deisobut | deisobutanizer |
| Del R | Del Rio |
| Dela | Delaware |
| delv | delivery (ed) (ability) |
| delv pt | delivery point |
| demur | demurrage |
| dend | dendrite (ic) |
| DENL | density log |
| depl | depletion |
| Depr | depreciation |
| deprop | depropanizer |
| dept | department |
| Des Crk | Desert Creek |
| Des M | Des Moines |
| desalt | deslater |
| desc | description |
| desorb | desorbent |
| desulf | desulfurizer |
| det | detail (s) |
| det | detector |
| deterg | detergent |
| detr | detrital |
| dev | deviate, deviation |
| Dev | Devonian |

| | |
|---|---|
| devel | develop (ed) (ment) |
| dewax | dewaxing |
| Dext | Dexter |
| DF | derrick floor |
| DF | diesel fuel |
| DF | drill floor |
| DFE | derrick floor elevation |
| DFO | datum faulted out |
| DFP | date of first production |
| DFT | dry film thickness |
| dg | decigram |
| DG | development gas well |
| DG | draft gauge |
| DG | dry gas |
| DGA | diglycolamine |
| DGTL | digital |
| DH | development well, helium |
| DH | double hub |
| DHC | dry-hole contribution |
| DHC | dry-hole cost |
| DHDD | dry-hole drilled deeper |
| DHDS | diesel hydrogen desulfurization |
| DHM | dry-hole money |
| DHR | dry-hole reentered |
| DHV | deemed heating value |
| dia | diameter |
| diag | diagonal |
| diag | diagram |
| diaph | diaphragm |
| dichlor | dichloride |
| diethyl | diethylene |
| diff | different (ial) (ence) |
| DIFL | dual injection focus log |
| dilut | diluted |

## Abbreviations with Definitions

| | |
|---|---|
| dim | dimension |
| dim | diminish (ing) |
| Din | Dinwoody |
| dir | direct (tion) (tor) |
| dir drlg | directional drilling |
| dir sur | directional survey |
| Disc | Discorbis |
| disc | discount |
| disc | discover (ed) (ing) (ion) |
| Disc. grav. | Discorbis gravelli |
| Disc. norm. | Discorbis normada |
| Disc. y. | Discorbis yeguaensis |
| disch | discharge |
| dism | disseminated |
| disman | dismantle |
| displ | displaced, displacement |
| dist | distance |
| dist | distillate, distillation |
| dist | district |
| distr | distribute (ed) (ing) (ion) |
| div | division |
| dk | dark |
| Dk Crk | Duck Creek |
| dl | deciliter |
| DL | drilling line |
| DLC | dual lower tubing |
| dlr | dealer |
| DLS | dogleg severity |
| DLT | dual lower tubing |
| DM | datum |
| dm | decimeter |
| DM | demand meter |
| DM | dipmeter |
| DM | drilling mud |

| | |
|---|---|
| dml | demolition |
| DMPD | dumped |
| dmpr | damper |
| DMS | dimethyl sulfide |
| dm | cubic decimeter |
| dm/s | cubic decimeter per second |
| dn | down |
| dns | dense |
| DO | development oil |
| DO | development oil well |
| do | ditto |
| DO | drill (ed) (ing) out |
| DOBLDG | dock operating building |
| DOC | diesel oil well |
| Doc | Dockum |
| doc | document |
| DOC | drilled-out cement |
| doc-tr | doctor-treatment |
| DOCREQ | Document request |
| DOD | drilled-out depth |
| DOE | Department of Energy |
| dolo | dolomite (ic) |
| dolst | dolstone |
| dom | domestic |
| dom AL | domestic airline |
| DOM WTR | domestic water |
| DOP | drilled-out plug |
| Dorn H | Dornick Hills |
| Doth | Dothan |
| Doug | Douglas |
| doz | dozen |
| DP | data processing |
| DP | dewpoint |
| DP | double pipe |

## Abbreviations with Definitions

| | |
|---|---|
| DP | drillpipe |
| DP SW | double pole switch |
| DPDB | double pole double base (switch) |
| DPDT SW | double pole double throw switch |
| dpg | deepening |
| DPM | drillpipe measurement |
| dpn | deepen |
| DBSB | double pole single base (switch) |
| DPST SW | double pole single throw switch |
| DPT | deep pool test |
| dpt | depth |
| dpt rec | depth recorder |
| DPU | drillpipe unloaded |
| DR | development redrill (sidetrack) |
| dr | drain |
| dr | drive |
| dr | drum |
| dr | druse |
| Dr Crk | Dry Creek |
| drk | derrick |
| DRL | double random lengths |
| drl | drill |
| drld | drilled |
| drlg | drilling |
| drlr | driller |
| DRM | drum |
| drng | drainage |
| dropd | dropped |
| drsy | drusy |
| DRV (R) | drive (ing) (er) |
| dry | drier, drying |
| ds | dense |
| DS | directional survey |
| DS | drillsite |

| | |
|---|---|
| DS | drillstem |
| DSF | drillsite facility |
| dsgn | design |
| DSI | drilling suspended indefinitely |
| dsl | diesel (oil) |
| dsmt (g) | dismantle (ing) |
| DSO | dead oil show |
| DSS | days since spudded |
| DST | drillstem test |
| DST (Strd) | drillstem test with straddle packers |
| dstl | distillate |
| dstn | destination |
| DSU | development well, sulfur |
| DSUPHTR | desuperheater |
| DT | downthrown |
| DT | drilling time |
| DTD | driller's total depth |
| DTH | decatherm |
| dtr | detrital |
| DTW | dealer tank wagon |
| DUC | dual upper casing |
| Dup | Duperow |
| dup | duplicate |
| DUR | duration |
| DUT | dual upper tubing |
| Dutch | Dutcher |
| DV | differential valve (cementing) |
| DVL | develop |
| DVT | davit |
| DWA | drilling with air |
| DWC | drilling and well completion |
| DWD | dirty water disposal |
| DWG | drawing |
| DWG | drilling with gas |

## Abbreviations with Definitions

| | |
|---|---|
| dwks | drawworks |
| DWM | drilling with mud |
| DWO | drilling with oil |
| DWP | dual (double) wall packer |
| DWSW | drilling with salt water |
| DWT | deadweight tester |
| DWT | deadweight tons |
| DWTR | dewatering |
| DX | development well workover |
| dx | duplex |
| dyn | dynamic |

| | |
|---|---|
| E | east |
| E | exploratory |
| E of W/L | east of west line |
| E.D. | effective depth |
| e.g. | for example |
| E.T.D | estimated total depth |
| E/BL | east boundary line |
| E/E | end to end |
| E/L | east line |
| E/O | east offset |
| E/2 | east half |
| E/4 | east quarter |
| ea | each |
| EA | environmental assessment |
| EAM | electric accounting machines |
| Earls | Earlsboro |
| Eau Clr | Eau Claire |
| ECC | eccentric |
| Ech | echinoid |

| | |
|---|---|
| ECM | East Cimarron Meridian (Oklahoma) |
| Econ | economics, economy, economizer |
| Ect | Ector (County, TX) |
| Ed lm | Edwards lime |
| EDC | ethylene dichloride |
| EDCTN(R) | education (tor) |
| EDD | expected date of delivery |
| EDP | electronic data processing |
| Educ | education |
| Edw | Edwards |
| EF | Eagle Ford |
| eff | effective |
| eff | efficiency |
| effl | effluent |
| EFV | equilibrium flash vaporization |
| Egl | Eagle |
| Eglwd | Englewood |
| EHP | effective horsepower |
| EIA | environmental assessment |
| EIR | environmental impact report |
| EIS | environmental impact statement |
| EJ | perforating, enerjet |
| EL | elevation (height) |
| ejtr | ejector |
| el gr | elevation ground |
| EL/T | electronic log tops |
| Elb | Elbert |
| ELB | elbow |
| elec | electric (al) |
| Elec/MAG | electromagnetic |
| elem | element, elementary |
| elev | elevation, elevator |
| Elg | Elgin |
| ELIM | eliminate (tor) (ed) |

| | |
|---|---|
| ell(s) | elbow (s) |
| Ellen | Ellenburger |
| Elm | Elmont |
| E'ly | easterly |
| EM | Eagle Mills |
| Emb | Embar |
| emer | emergency |
| EMF | electromotive force |
| EMN | electromagnetic |
| EMNI | electromagnetic induction |
| EMP | European melting point |
| empl | employee |
| EMS | Ellis-Madison contact |
| emul | emulsion |
| encl | enclosure |
| End | Endicott |
| endo | endothyra |
| eng | engine |
| engr (g) | engineer (ing) |
| enl | enlarged |
| enml | enamel |
| Ent | Entrada |
| ENT | entrance |
| ent | entry |
| ENV | envelope |
| ENVIR | environment |
| EO | emergency order |
| Eoc | Eocene |
| EOF | end of file |
| EOL | end of line |
| EOM | end of month |
| EOQ | end of quarter |
| EOR | east of Rockies |
| EOR | enhanced oil recovery |

| | |
|---|---|
| EOY | end of year |
| EP | end point |
| EP | extreme pressure |
| Epon. | Eponides |
| Epon. y. | Eponides yeguaensis |
| eq | equal, equalizer |
| Eq | equation (before a number) |
| equip | equipment |
| equiv | equivalent |
| ERC/MK | erection mark |
| erect | erection |
| Eric | Ericson |
| ERW | electric resistance weld |
| ESD | emergency shutdown |
| est | estate |
| est | estimate (ed) (ing) |
| et al. | and others |
| et con. | and husband |
| et seq. | and the following |
| et ux. | and wife |
| et vir. | and husband |
| ETA | estimated time of arrival |
| eth | ethane |
| ethyle | ethylene |
| EU | Eutaw |
| EUE | external upset end |
| euhed | euhedral |
| EUR | estimated ultimate recovery |
| ev | electron-volts |
| ev-sort | even sorted |
| eval | evaluate |
| evap | evaporation, evaporate |
| EW | electric weld |
| EW | exploratory well |

# Abbreviations with Definitions

| | |
|---|---|
| EWT | early well tie-ins |
| EX | example |
| ex | except |
| Ex | Exeter |
| EX-PRF | explosion proof |
| EXAM | Examination |
| EXC | Excitation |
| exc | Excavation |
| exch | Exchanger |
| excl | Excellent |
| EXEC | Executive |
| exh | Exhaust |
| exh | exhibit |
| exist | existing |
| EXP | expansion |
| exp | expense |
| EXP JT | expansion joint |
| exp plg | expendable plug |
| expir | expire (ed) (ing) (ation) |
| expl | exploratory, exploration |
| explos | explosive |
| exr | executor |
| Exrx | executrix |
| exst | existing |
| ext | external |
| Ext M/H | extension manhole |
| ext (n) | extended, extension |
| extr | exterior |
| extrac | extraction |
| EYC | estimated yearly consumption |

## F

| | |
|---|---|
| F&D | faced and drilled |
| F&D | flanged and dished (heads) |
| F&F | fuels & fabrication |
| F&L | fuels & lubricants |
| F&S | flanged and spigot |
| F to F | face to face |
| F.G. | fracture gratient |
| F.O.E | fuel oil equivalent |
| F-D | formation density |
| F-DIA | flow diagram |
| f-gr | fine grained |
| F-MET | flowmeter |
| F-R oil | fire-resistant oil |
| F-SHT | flow sheet |
| F/ | flowed, flowing |
| F/DIA | flow diagram |
| F/FAB | field fabricated |
| F/GOR | formation gas-oil ratio |
| F/O opt | farmout option |
| F/S | flange & screwed |
| F/S | front & side |
| F/SW | flow switch |
| F/WTR | fire water |
| f/xin | finely crystalline |
| fab | fabricate (ed) (tion) |
| FAB | faint air blow |
| fac | facet (ed) |
| FACIL | facility (ies) |
| FACO | field authorized to commence operations |
| fail | failure |

## Abbreviations with Definitions

| | |
|---|---|
| Fall Riv | Fall River |
| FAO | finish all over |
| Farm | Farmington |
| FARO | flowed (ing) at the rate of |
| fau | fauna |
| FB | fresh break |
| FBH | flowing by heads |
| FBHP | flowing bottom-hole pressure |
| FBHPF | final bottom-hole pressure, flowing |
| FBHPSI | final bottom-hole pressure, shut-in |
| FBP | final boiling point |
| fbrgs | fiberglass |
| FC | filter cake |
| FC | fixed carbon |
| FC | float collar |
| FCC | fluid catalytic cracking |
| FCL | facility capacity limits |
| FCO | functional check out |
| FCP | flowing casing pressure |
| FCV | flow control valve |
| FD | feed |
| FD | floor drain |
| FD | flow diagram |
| FD EFF | feed effluent |
| FD/WTR | feed water |
| FDC | formation density correlated |
| FDL | formation density log |
| fdn | foundation |
| fdr | feeder |
| Fe | iron |
| Fe-st | ironstone |
| FE/L | from east line |
| fed | federal |
| FEIS | Final Environmental Impact Statement |

| | |
|---|---|
| FEL | from east line |
| FELA | Federal Employers Liability Act |
| FEM | female |
| Ferg | Ferguson |
| ferr | ferruginous |
| fert | fertilizer |
| $Fe_2(SO_4)_3$ | ferric sulfate |
| FF | fishing for |
| FF | flat face |
| FF | fracfinder (log) |
| FF | full of fluid |
| FFA | female to female angle |
| FFA | full freight allowed (purchasing term) |
| FFG | female to female globe (valve) |
| FFGU | field fuel gas unit |
| FFL | final fluid level |
| FFO | furnace fuel oil |
| FFP | final flowing pressure |
| FG | fuel gas |
| FGIH | finish going in hole |
| FGIW | finish going in with |
| FGVV | flanged gate valve |
| FH | full hole |
| FHP | final hydrostatic pressure |
| FI | flow indicator |
| fib | fibrous |
| FIC | flow-indicating controller |
| fig | figure |
| FIH | finished in hole |
| FIH | fluid in hole |
| filt | filtrate |
| fin | final |
| fin | finish (ed) |
| fin drlg | finished drilling |

## Abbreviations with Definitions

| | |
|---|---|
| FIN GR | finish grade |
| FIRC | flow indicating ratio controller |
| fis | fissure |
| fish | fishing |
| fisl | fissile |
| FIT | formation interval tester |
| fix | fixture |
| FJ | flush joint |
| Flt | fault |
| FL | flashing |
| FL | floor |
| FL | flow line |
| fl | fluid |
| FL | fluid level |
| FL | flush |
| Fl-COC | flash point, Cleveland Open Cup |
| fl/ | flowed (ing) |
| FL/BD | flammable liquid building |
| FLA | Ferry Lake anhydrite |
| flat | flattened |
| Flath | Flathead |
| fld | failed |
| fld | feldspar (thic) |
| fld | field |
| FLD | full-length drift |
| flex | flexible |
| flg | flowing |
| flg (d) (s) | flange (ed) (es) |
| Flip | Flippen |
| flk | flaky |
| FLMB | flammable |
| flo | flow |
| FLO | flushing oil |
| Flor fl | Florence flint |

| | |
|---|---|
| flshd | flushed |
| flt | float |
| fltg | floating |
| fltn | flotation |
| FLTR | filter |
| flu | flue |
| Flu | fluid |
| fluor | fluorescence, fluorescent |
| flw (d) (g) | flow (ed) (ing) |
| Flwg Pr. | flowing pressure |
| Flwrpt | flowerpot |
| FLXBX | flexibox |
| fm | formation |
| FM | frequency meter |
| FM | frequency modulation |
| Fm W | formation water |
| f'man | foreman |
| Fpm | feet per minute |
| Fn | fine |
| FNEL | from northeast line |
| FNL | from north line |
| Fnly | finely |
| FNSH | finish |
| Fnt | faint |
| FNWL | from northwest line |
| FO | farmout |
| FO | faulted out |
| FO | final open |
| FO | fuel oil |
| FO | full opening |
| FOB | free on board |
| FOCL | focused log |
| FOE-WOE | flanged one end, welded one end |
| FOH | full open head |

## Abbreviations with Definitions

| | |
|---|---|
| Fol | foliated |
| FONSI | finding of no significant impact |
| FOR | fuel oil return |
| Forak | Foraker |
| Foram | foraminifera |
| Fort | Fortura |
| FOS | face of stud |
| FOS | fuel oil supply |
| Foss | fossiliferous |
| FOT | flowing on test |
| Fount | Fountain |
| Fox H | Fox Hills |
| FP | final pressure |
| FP | flowing pressure |
| FP | freezing point |
| Fph | Feet per hour |
| FPI | free-point indicator |
| FPO | field purchase order |
| Fprf | fireproof |
| Fps | feet per second |
| f-p-s | foot-pound-second (system) |
| FPSO | floating production storage and offloading vessel |
| FPT | female pipe thread |
| FPTFD | field pressure test flow diagram |
| FQG | frosted quartz grains |
| fr | fair |
| FR | feed rate |
| FR | flow rate |
| FR | flow recorder |
| fr | fractional |
| fr | from |
| fr | front |
| fr | frosted |
| fr E/L | from east line |

| | |
|---|---|
| fr N/L | from north line |
| fr S/L | from south line |
| fr W/L | from west line |
| FRA | friction reducing agent |
| frac (d) (s) | fracture, fractured, fractures |
| fract | fractionation, fractionator, fractional |
| frag | fragment |
| fran | franchise |
| Franc | Franconia |
| FRC | flow recorder control |
| Fred | Fredericksburg |
| Fred | Fredonia |
| freq | frequency |
| FRG | forge (ed) (ing) |
| Frgy | froggy |
| fri | friable |
| FRM (G) | frame, framing |
| Fron | frontier |
| fros | frosted |
| FRP | fiberglass-reinforced plastic |
| FRR | field receiving report |
| FRR | final report for rig |
| frs | fresh |
| frt | freight |
| Fruit | Fruitland |
| FRW | final report for well |
| frwk | framework |
| frzr | freezer |
| FS | feedstock |
| FS | float shoe |
| FS | flow station |
| FS | forged steel |
| FS&WLs | from south and west lines |
| FSEL | from southeast line |

## Abbreviations with Definitions

| | |
|---|---|
| fsg | fishing |
| FSIP | final shutin pressure |
| FSL | from south line |
| FSP | flowing surface pressure |
| FST | forged steel |
| FSTN | fasten (ing) (er) |
| FSWL | from southwest line |
| ft | feet, foot |
| FT | firm transport |
| FT | formation test |
| Ft C | Fort Chadborne |
| Ft H | Fort Hayes |
| ft-lb | foot-pound |
| ft-lb/hr | foot-pound per hour |
| Ft R | Fort Riley |
| Ft U | Fort Union |
| Ft W | Fort Worth |
| ft-c | foot-candle |
| ftg | fittings |
| ftg | footing, footage |
| FTP | final (flowing) tubing pressure |
| FTS | fluid to surface |
| $ft^2$ | square feet |
| $ft^3$ | cubic feet |
| FU | fill up |
| Full | Fullerton |
| furf | furfural |
| furn | furnace |
| FURN | furnish (ed) |
| Furn & fix | furniture and fixtures |
| Fus | Fuson |
| Fussel | Fusselman |
| Fusul | Fusulinid |
| fut | future |

| | |
|---|---|
| FV | funnel viscosity |
| fvst | favosites |
| FW | fillet weld |
| FW | fresh water |
| FWC | field wildcat |
| fwd | forward |
| FWD | four-wheel drive |
| FWL | from west line |
| fwtr | fresh water |
| fxd | fixed |
| FYE | fiscal year ending |
| FYI | for your information |

## G

| | |
|---|---|
| G | gas |
| G egg | goose egg |
| g mole | gram molecular weight |
| G Rk | gas rock |
| G.M. | gravity meter |
| G. Riv | Gull River |
| g-cal | gram calorie |
| $G-N_2$ | gaseous nitrogen |
| $G-O_2$ | gaseous oxygen |
| G&MCO | gas & mud-cut oil |
| G&O | gas and oil |
| G&OCM | gas- and oil-cut mud |
| G/L | gathering line |
| G/P | gun perforate |
| ga | gauge (ed) (ing) |
| GA | gallons acid |
| GA | general agreement |
| GAF | gross acre-feet |

## Abbreviations with Definitions

| | |
|---|---|
| gal | gallon(s) |
| gal sol | gallons of solution |
| gal/Mcf | gallons per thousand cubic feet |
| gal/min | gallons per minute |
| Gall | Gallatin |
| galv | galvanized |
| gaso | gasoline |
| gast | gastropod |
| GB | gun barrel |
| GBDA | gallons breakdown acid |
| GC | gas-cut |
| GCAW | gas-cut acid water |
| GCD | gas-cut distillate |
| GCLO | gas-cut load oil |
| GCLW | gas-cut load water |
| GCM | gas-cut mud |
| GCO | gas-cut oil |
| GCPA | gas cap participating area |
| GCPD | gallons condensate per day |
| GCPH | gallons condensate per hour |
| GCR | gas-condensate ratio |
| GCSW | gas-cut salt water |
| GCT | guidance continuance tool |
| GCW | gas-cut water |
| GD li | Glen Dean lime |
| gd | good |
| gd o&t | good odor & taste |
| GDE | geothermal development, failure |
| Gdld | Goodland |
| GDR | gas-distillate ratio |
| GDS | geothermal development, success |
| Gdwn | Goodwin |
| GE | General Electric Company |
| GE | grooved ends |

| | |
|---|---|
| gel | gelled |
| gel | jelly-like colloidal suspension |
| gen | generation, generator |
| genl | general |
| Geo | Georgetown |
| geo | geothermal |
| geol | geology (ist) (ical) |
| geop | geophysics (ical) |
| GFLU | good fluorescence |
| GFR | gas-fluid ratio |
| gg | grains per gallon |
| GGD | gas lift gas distribution |
| GGW | gallons gelled water |
| GH | Greenhorn |
| GHO | gallons heavy oil |
| GHSG | gas-handling study group |
| GI | gas injection |
| Gib | Gibson |
| GIH | going in hole |
| gil | gilsonite |
| Gilc | Gilcrease |
| GIP | gas in pipe |
| GISB | gas industry standards board |
| GIW | gas injection well |
| GJ | gigajoule |
| GL | gas lift |
| GL | ground level |
| glau | glauconite, glauconitic |
| GLBVV | globe valve |
| gld thd | galled threads |
| Glen | Glenwood |
| glna | Galena |
| GLO | gas lift oil |
| GLO | General Land Office (Texas) |

## Abbreviations with Definitions

| | |
|---|---|
| Glob | Globigerina |
| Glor | Glorieta |
| GLR | gas-liquid ratio |
| gls (y) | glass, glassy |
| GLT | gas lift transfer |
| glyc | glycol |
| GMC | General Motors Corporation |
| gm | gram |
| GM | ground measurement (elevation) |
| gm-cal | gram-calorie |
| GMA | gallons mud acid |
| gmy | gummy |
| gnd | grained (as in fine-grained) |
| gns | gneiss |
| GO | gallons oil |
| GO | gas odor |
| GO | grind out |
| GOC | gas-oil contact |
| GODT | gas odor distillate taste |
| Gol | Golconda lime |
| Good L | Goodland |
| GOPD | gallons of oil per day |
| GOPH | gallons of oil per hour |
| GOR | gas-oil ratio |
| Gor | Gorham |
| Gouldb | Gouldbusk |
| gov | governor |
| govt | government |
| GP | gas pay |
| GP | gasoline plant |
| GPC | gas purchase contract |
| GPD | gallons per day |
| GPF | granite point field |
| GPG | grains per gallon |

| | |
|---|---|
| GPH | gallons per hour |
| GPM | gallons per minute |
| GPM | geophysical investigation map |
| GPS | gallons per second |
| GPU | Gas Production Unit |
| GQM | geological quadrangle map |
| GR | gamma ray |
| GR | gauge ring |
| GR | Glen Rose |
| gr | grade |
| gr | grain |
| gr | gravity |
| gr | grease |
| gr | ground |
| gr API | gravityo API |
| gr roy | green royalty |
| Gr Sd | Gray sand |
| gr wt | gross weight |
| GR&DC | Gulf Research and Development Company |
| GRA | gallons regular acid |
| GRAD | gradiomanometer |
| grad | gradual, gradually |
| gran | granite, granular |
| gran w | granite wash |
| Granos | Graneros |
| grap | graptolite |
| grav | gravity |
| Gray | Grayson |
| Grayb | Grayburg |
| grdg | grading |
| grdg loc | grading location |
| GRDL | guard log |
| GRG | gas reserve group |
| GRI | gas research institute |

## Abbreviations with Definitions

| | |
|---|---|
| grn | green |
| Grn Riv | Green River |
| grn sh | green shale |
| grnd | ground |
| grnlr | granular |
| GRP | group |
| GRS | gas to surface |
| grs | gross |
| grt | grant (of land) |
| grtg | grating |
| grty | gritty |
| grv | grooved |
| grvt | gravitometer |
| GRVTY | gravity |
| gry | gray |
| GS | gas show |
| GS | guide shoe |
| GSC | gas sales contract |
| GSG | good show of gas |
| GWSI | gas well shutin |
| gskt | gasket |
| GSO | good show oil |
| GSO&G | good show oil and gas |
| GST | gamma spectroscopy tool |
| GSW | gallons salt water |
| gsy | greasy |
| GT | geothermal |
| GTS | gas to surface (time) |
| GTSTM | gas too small to measure |
| GU | gas unit |
| Guns | Gunsite |
| GUS | gusset |
| GV | gas volume |
| gvl | gravel |

Standard Oil & Gas Abbreviator

| | |
|---|---|
| GVLPK | gravel packet |
| GVNM | gas volume not measured |
| GW | gallons water |
| GW | gas well |
| GW | geothermal wildcat, failure |
| GWC | gas-water contact |
| GWD | geothermal wildcat, success |
| GWG | gas-well gas |
| GWPH | gallons of water per hour |
| gyp | gypsum |
| gypy | gypsiferous |
| Gyr. | Gyroidina |
| Gyr. sc. | Gyroidina scal |
| gywk | greywacke |

## H

| | |
|---|---|
| H&V | heating and ventilating |
| H H P | hydraulic horsepower |
| H. O. | hole opener |
| H-SEL | perforating hyper select |
| H-VOLT | high voltage |
| H/C | hydrocracker |
| Hackb | Hackberry |
| Hara | Haragan |
| Hask | Haskell |
| Haynes | Haynesville |
| haz | hazardous |
| HB | house brand (regular grade of gasoline) |
| Hburg | Hardinsburg sand (local) |
| HBP | held by production |
| hbr | harbor |
| HC | hydrocarbon |

## Abbreviations with Definitions

| | |
|---|---|
| HC | hydrocracker |
| HCDS | hydrocarbon drain system |
| HCGO | heavy coker gas oil |
| HCO | heavy cycle oil |
| HCV | hand-control valve |
| hd | hard |
| hd | head |
| HD | heavy duty |
| HD | high detergent |
| HD | hot dry rock development, failure |
| HD | Hydril |
| HD | perforating, hyperdome |
| Hd li | hard lime |
| hd sd | hard sand |
| hdl | handle |
| hdns | hardness |
| hdr | header |
| HDS | hot dry rock development, successful |
| HDS | hydrogen delsulfurization |
| Hdwr | hardware |
| Heeb | Heebner |
| Hem | hematite |
| Her | Herington |
| Herm | Hermosa |
| Het | heterostegina |
| HEX | heat exchanger |
| hex | hexagon (al) |
| hex | hexane |
| HEX HD | hex head |
| hfg | hydrofining |
| HFO | heavy fuel oil |
| HFO | hole full of oil |
| HFSW | hole full of salt water |
| HFW | hole full of water |

| | |
|---|---|
| HGCM | heavily (highly) gas-cut mud |
| HGCSW | heavily (highly) gas-cut salt water |
| HGCW | heavily (highly) gas-cut water |
| HGOR | high gas-oil ratio |
| hgr | hanger |
| hgt | height |
| HH | hand hole |
| HH | hydrostatic heat |
| HIA | Hydrologic Investigations Atlas |
| Hick | Hickory |
| Hill | Hilliard |
| hky | hackly |
| HLDN | hold down |
| HLSD | high-level shutdown |
| HND/WHL | handwheel |
| HNDLG | handling |
| HO | heating oil |
| HO | heavy oil |
| HO | hole opener |
| HO | home office |
| HO&GCM | heavily (highly) oil- and gas-cut mud |
| hock | hockleyensis |
| HOCM | heavily (highly) oil-cut mud |
| HOCSW | heavily (highly) oil-cut salt water |
| HOCW | heavily (highly) oil-cut water |
| Hog | Hogshooter |
| Holl | Hollandberg |
| Home Cr | Home Creek |
| hop | hooper |
| horiz | horizontal |
| Hodp | Hospah |
| HOT | hot oil tar |
| Hov | Hoover |
| Hox | Hoxbar |

## Abbreviations with Definitions

| | |
|---|---|
| HP | high pressure |
| HP | horsepower |
| HP | hydraulic pump |
| HP | hydrostatic pressure |
| hp-hr | horsepower-hour |
| HPF | holes per foot |
| HPG | high-pressure gas |
| HPG | high-pressure gauge |
| HQ | headquarters |
| HR | heavy reformate |
| hr | hour (s) |
| HR Sul W | hole full of sulfur water |
| HRD | high-resolution dipmeter |
| hrs | heirs |
| HRS | hot-rolled steel |
| HSD | heavy steel drum |
| HSE | house (ed) (ing) |
| HST | hydrostatic test |
| HT | heat tracing (ed) |
| HT | heat-treated, heater treater |
| HT | high temperature |
| HT | high tension |
| HTA | heat-treated alloy |
| htr | heater |
| HTSD | high-temperature shutdown |
| HU | hook up |
| Humb | Humblei |
| Hump | Humphreys |
| Hun | Hunton |
| HUX | heavy hydrocrackate |
| HV | high viscosity |
| HVAC | heating ventilating and air conditioning |
| HVGO | heavy gas oil |
| HVI | high viscosity index |

| | |
|---|---|
| HVL | high volume lift |
| hvly | heavily |
| hvy | heavy |
| HW | hot dry rock wildcat, failure |
| HWCM | heavily (highly) water-cut mud |
| HWD | hot dry rock wildcat, success |
| HWP | hookwall packer |
| hwy | highway |
| HX | heat exchanger |
| HYD | hydraulic |
| HYD | Hydril thread |
| HYDA | Hydril Type A joint |
| HYDCA | Hydril Type CA joint |
| HYDCS | Hydril Type CS Joint |
| HYDRO | hydro test |
| hydtr | hydrotreater |
| Hyg | hygiene |
| HYGN | hydrogenation |
| HYPO | hypotenuse |
| Hz | Hertz |
| $H_2$ | hydrogen |
| $H_2S$ | hydrogen sulfite |
| $H_2SO_4$ | sulfuric acid |

## I

| | |
|---|---|
| I.D sign | identification sign |
| I- | miscellaneous investigations series |
| I-O-M | installation operation & maintenance |
| I/C | interconnection (ing) |
| I/CFD | interconnecting flow diagram |
| I/O | input/output |
| IAB | initial air block |

## Abbreviations with Definitions

| | |
|---|---|
| IB | impression block |
| IB | iron body (valve) |
| IBBC | iron body brass core (valve) |
| IBBM | iron body brass (bronze) mounted (valve) |
| IBHP | initial bottom-hole pressure |
| IBHPF | initial bottom-hole pressure, flowing |
| IBHPSI | initial bottom-hole pressure, shut in |
| IBP | initial boiling point |
| IC | iron case |
| icfos | microfossil (iferous) |
| ID | inside diameter |
| IDENT | identify (ier) (ication) |
| Idio | Idiomorpha |
| IES | induction electrical survey |
| IF | internal flush |
| IFL | initial fluid level |
| IFP | initial flowing pressure |
| IG | injection gas |
| Ign | igneous |
| IGN | ignition |
| IGOR | injection gas-oil ratio |
| IGV/IBV | inlet gate valve/inlet ball valve |
| IH | in hole |
| IHP | indicated horsepower |
| IHP | initial hydrostatic pressure |
| IHPHR | indicated horsepower hour |
| II | injection index |
| IJ | integral joint |
| ILUM | illuminator (s) |
| imdb | imbedded |
| IMF | intermediate manifolds |
| immed | immediate (ly) |
| Imp | Imperial |
| IMP | impounding |

| | |
|---|---|
| Imp gal | Imperial gallon |
| imperv | impervious |
| IMW | initial mud weight |
| in. | inch (es) |
| in. Hg | inches mercury |
| in.-lb | inch-pound |
| in./sec | inches per second |
| Inbded | imbedded |
| Inc. | incorporated |
| inc | increment |
| incd | incandescent |
| INCIN | incinerator, incineration |
| incl | include (ed) (ing) |
| INCLR | intercooler |
| incls | inclusions |
| INCM | income (er) (ing) |
| incr | increase (ed) (ing) |
| ind | induction |
| indic | indicate (s) (tion) |
| indiv | individual |
| indr | indurated |
| indst | indistinct |
| Inf. L | inflammable liquid |
| Inf. S | inflammable solid |
| info | information |
| ingr | intergranular |
| inhib | inhibitor |
| init | initial |
| Inj | injection, injected |
| Inj Pr | injection pressure |
| Inl | inland |
| Inl | inlet |
| Inland | interlaminated |
| Inoc | Inoceramus |

## Abbreviations with Definitions

| | |
|---|---|
| INPE | installing (ed) pumping equipment |
| INQ | inquire, inquiry |
| ins | insulate, insulation |
| ins | insurance |
| insole | insoluble |
| inps | inspect (ed) (ing) (tion) |
| inst (d) (g) (l) | install (ed) (ing) (tion) |
| inst | instantaneous |
| inst | institute |
| instr | instrument, instrumentation |
| insul | insulate |
| INT | integral |
| int | interest |
| int | interior |
| int | internal |
| int | intersection |
| intclr | intercooler |
| INTCON | interconnection |
| inter-gran | intergranular |
| inter-lam | interlaminated |
| inter-xln | intercrystalline |
| interbd | interbedded |
| intgr | integrator |
| INTL | internal |
| intl | interstitial |
| intr | intrusion |
| ints | intersect |
| intv | interval |
| inv | invert (ed) |
| inv | invoice |
| invrtb | invertebrate |
| IO | initial open |
| IP | initial potential |
| IP | initial pressure |

65

| | |
|---|---|
| IP | initial production |
| IP | intermediate pressure |
| IPA | initial participating area |
| IPA | isopropyl alcohol |
| IPE | install (ing) pumping equipment |
| IPF | initial production flowed (ing) |
| IPG | initial production gas lift |
| IPI | initial production on intermitter |
| IPL | initial production plunger lift |
| IPP | initial production pumping |
| IPR | inflow performance rate |
| IPS | initial production swabbing |
| IPS | iron pipe size |
| IPT | iron pipe thread |
| IR | infrared |
| IR | injection rate |
| Ire | Ireton |
| irid | iridescent |
| irreg | irregular |
| IRS | Internal Revenue Service |
| Irst | ironstone |
| IS | inside screw (valve) |
| ISIP | initial shutin pressure (DST) |
| ISIP | instantaneous shutin pressure (frac) |
| ISITP | initial shutin tubing pressure |
| ISO | isometric |
| ISO/CKR | isocracker |
| ISOL | isolate (tor) |
| isom | isometric |
| isoth | isothermal |
| ISS | issue |
| IT | interruptible transport |
| ITB | invitation to bid |
| ITC | investment tax credit |

## Abbreviations with Definitions

| | |
|---|---|
| ITD | intention to drill |
| IUE | internal upset ends |
| Ives | Iverson |
| IVP | initial vapor pressure |
| IW | injection water |
| IW | injection well |

## J

| | |
|---|---|
| J | Joule |
| J&A | junked and abandoned |
| J/O | joint operation |
| jac | jacket |
| Jack | Jackson |
| Jasp | jasper (oid) |
| Jax | Jackson sand |
| JB | junction box |
| JB | junk basket |
| jbr | jobber |
| JC | job complete |
| jct | junction |
| Jdn | Jordan |
| Jeff | Jefferson |
| JFA | jet fuel (aviation) |
| JIB | joint interesting billing |
| JINO | joint interest nonoperated (property) |
| JJ | junk joint |
| jmd | jammed |
| jnk | junk (ed) |
| JOA | joint operating agreement |
| JOP | joint operating provisions |
| JP | jet perforated |
| JP fuel | jet propulsion fuel |

| | |
|---|---|
| JP/ft | jet perforations per foot |
| JSPF | jet shots per foot |
| jt(s) | joint (s) |
| JTU | jet treating unit |
| Jud Riv | Judith River |
| Jur | Jurassic |
| juris | jurisdiction |
| JV | joint venture |
| Jxn | Jackson |

## K

| | |
|---|---|
| K | Kelvin (temperature scale) |
| K | thousand (i.e., 13K = 13,000) |
| Kai | Kaibib |
| kao | kaolin |
| Kay | Kayenta |
| KB | kelly bushing |
| KBM | kelly bushing measurement |
| KC | Kansas City |
| kc | kilocycle |
| kcal | kilocalorie |
| KD | kiln dried |
| KD | Kincaid lime |
| KD | knock down |
| KDB | kelly drive bushing |
| KDB-LDG FLG | kelly drill bushing to landing flange |
| KDB-MLW | kelly drill bushing to mean low water |
| KDB-Plat | kelly drill bushing to platform |
| KDBE | kelly drive bushing elevation |
| Ke | Keener |
| Keo-Bur | Keokuk-Burlington |
| kero | kerosine (kerosene) |

## Abbreviations with Definitions

| | |
|---|---|
| ket | ketone |
| KEV | thousand electron-volts |
| Key | Keystone |
| kg | kilogram |
| kg-cal | kilogram-calorie |
| kg-m | kilogram-meter |
| KGRA | known geothermal resource area |
| Khk | Kinderhook |
| kHz | kilohertz |
| Kia | Kiamichi |
| Kib | Kibbey |
| Kin li | Kincaid Lime |
| kin | kinematic |
| kip | one thousand pounds |
| kip-ft | one thousand foot-pounds |
| Kirt | Kirtland |
| KIT | kitchen |
| kl | kiloliter |
| kld | killed |
| km | kilometer |
| KMA | KMA sand |
| KO | kick off |
| KO | knock out |
| Koot | Kootenai |
| KOP | kickoff point |
| kPa | kilopascal |
| Kri | Krider |
| ktle | kettle |
| kv | kilovolt |
| KV | kinematic viscosity |
| KV | permeability (vertical direction) |
| kva | kilovolt-ampere |
| kvah | kilovolt-ampere-hour |
| kvar | kilovar; reactive kilovolt-ampere |

| | |
|---|---|
| kvar-hr | kilovar-hour |
| kvp | kilovolt peak |
| KW | kill (ed) well |
| kw | kilowatt |
| kwh | kilowatt-hour |
| kwhm | kilowatt-hour-meter |

## L

| | |
|---|---|
| l | liter |
| L&P | ladder & platform |
| L U | lease use (gas) |
| L.P. | line pipe |
| L-DK | loading dock |
| L-VOLT | low voltage |
| L/ | Lower, i.e., L/Gallup |
| L/Alb | Lower Albany |
| L/Cret | Lower Cretaceous |
| L/Tus | Lower Tuscaloosa |
| LA | level alarm |
| LA | lightning arrester |
| LA | load acid |
| La Mte | La Motte |
| lab | labor |
| lab | laboratory |
| LACT | lease automatic custody transfer |
| lad | ladder |
| LAG | lagging |
| Lak | Lakota |
| lam | laminated, lamination (s) |
| Land | Landulina |
| Lans | Lansing |
| Lar | Laramie |

# Abbreviations with Definitions

| | |
|---|---|
| LAS | lower anhydrite stringer |
| lat | latitude |
| Laud | Lauders |
| Layt | Layton |
| LB | light barrel |
| lb | pound |
| lb-in. | pound-inch |
| lb/ft | pounds per foot |
| lb/sq ft | pounds per square foot |
| LBOS | light brown oil stain |
| lbr | lumber |
| LC | lease crude |
| LC | level controller |
| LC | long coupling |
| LC | lost circulation |
| LC | lower casing |
| LC | lug cover type (5-gallon can) |
| LCGO | light coker gas oil |
| lchd | leached |
| LCL | less-than carload lot |
| LCL | local |
| LCM | lost circulation material |
| LCP | local control panel |
| LCP | lug cover wit pour spout |
| LCV | level control valve |
| LD | laid down |
| ld | load |
| ld(s) | land (s) |
| LDC | laid-down cost |
| LDC | local distribution company |
| LDCX | lead drill collar |
| LDDCs | laid (laying) down drill collars |
| LDDP | laid (laying) down drillpipe |
| LDFL | large-diameter flow line |

| | |
|---|---|
| LDG | landing |
| LDG | loading |
| LDR | loader |
| Le C | Le Comptom |
| Leadv | Leadville |
| LEL | lower explosive limit |
| Len | Lennep |
| len | lenticular |
| LFO | light fuel oil |
| lg | large |
| lg | length |
| lg | level glass |
| lg | long |
| Lg Disc | Large Discorbis |
| LGD | Lower Glen Dean |
| Lge | league |
| LH | left hand |
| LH/RP | long handle / round point |
| LI | level indicator |
| LIB | light iron barrel |
| LIC | level indicator controller |
| lic | license |
| Lieb | Liebuscella |
| lig | lignite, lignitic |
| LIGB | light iron grease barrel |
| LIH | left in hole |
| lim | limit, limonite |
| lin | linear |
| lin | liner |
| lin ft | linear foot |
| LIP | local injections plants |
| liq | liquid |
| liqftn | liquefaction |
| litho | lithographic |

## Abbreviations with Definitions

| | |
|---|---|
| LJ | lap joint |
| lk | leak |
| lk | lock |
| LKG | leakage |
| LKR | locker |
| LLC | liquid level controller |
| LLG | liquid level gauge |
| lm | lime, limestone |
| LMF | lowermost flange |
| LMn | Lower Menard |
| Lmpy | lumpy |
| LMTD | log mean temperature difference |
| lmy | limy |
| Lmy sh | limy shale |
| LN | line |
| ln | logarithm (natural) |
| LNG | liquefied natural gas |
| lngl | linguloid |
| lnr | liner |
| lns | lense |
| LO | load oil |
| LO | lube oil |
| LOA | length overall |
| loc | located, location |
| loc abnd | location abandoned |
| loc gr | location graded |
| log | logarithm (common) |
| long | longitude (inal) |
| LOS | lease operations |
| Lov | Lovell |
| Lov | Lovington |
| low | lower |
| LOX | liquid oxygen |
| LP | line pressure |

| | |
|---|---|
| LP | lodge pole |
| lp | loop |
| LP | low pressure |
| LP sep | low-pressure separator |
| LP-Gas | liquefied petroleum gas |
| LPG | liquefied petroleum gas |
| LPG | propane |
| LPO | local purchase order |
| LPS | low-pressure separation |
| LR | level recorder |
| LR | long radius |
| LRAP | long-range automation plan |
| LRC | level recorder controller |
| lrg | large |
| LRP | long-range plan |
| ls | limestone |
| LS | long string |
| LSD | legal subdivision (Canada) |
| LSD | light steel drum |
| lse | lease |
| LST/COMPTS | list of components |
| lstr | lustre |
| lt | light |
| LT | lower tubing |
| LT&C | long threads and coupling |
| ltd | limited |
| LTD | log total depth |
| ltg | lighting |
| LTL | less than truckload |
| ltl | little |
| ltr | letter |
| LTS Unit | low-temperature separation unit |
| LTSD | low-temperature shutdown |
| LTT | long-term tubing test |

## Abbreviations with Definitions

| | |
|---|---|
| LTX unit | low-temperature extraction unit |
| lub | lubricate (ed) (ing) (tion) |
| Lued | Lueders |
| LUG | lse use gas |
| LUX | light hytrocrackate |
| LV | liquid volume |
| LVI | low viscosity index |
| lvl | level |
| Lvnwth | Leavenworth |
| LW | lapweld |
| LW | load water |
| LWL | low water loss |
| lwr | lower |

## M

| | |
|---|---|
| M | mesh |
| m | meter |
| µ | micron |
| mµ | millimicron |
| M | molar |
| M | thousand (i.e., 9M = 9,000) |
| M&R Sta. | measuring and regulating station |
| M. Tus | Marine Tuscaloosa |
| µ-a | microampere |
| µ-f | microfarad |
| µ-g | microgram |
| m-gr | medium grained |
| µ-in. | microinch |
| m-kg | meter-kilogram |
| µ-µ | micromicron |
| µ-µf | micro-microfarad |
| µ-v | microvolt |

## Standard Oil & Gas Abbreviator

| | |
|---|---|
| M&F | male and female (joint) |
| M&FP | maximum & final pressure |
| M&P | mix and pump |
| M/ | middle |
| m/l | more or less |
| M/PLT | masking plate |
| M/T | marine terminal |
| M/V | motor vehicle, motor vessel |
| MA | massive anhydrite |
| ma | microampere |
| ma | milliampere |
| MA | mud acid |
| MAC | medium amber cut |
| mach | machine |
| Mack | Mackhank |
| Mad | Madison |
| mag | magnetic, magnetometer |
| maint | maintenance |
| maj | major, majority |
| mall | malleable |
| man | manifold |
| man | manual |
| man op | manually operated |
| Manit | Manitoban |
| Mann | Manning |
| MAOP | maximum allowable operating pressure |
| Maq | Maquoketa |
| mar | marine |
| mar | maroon |
| March | Marchand |
| marg | marginal |
| Marg. | Marginulina |
| Marg. coco | Marginulina coco |
| Marg. fl. | Marginulina flat |

| | |
|---|---|
| Marg. rd | Marginulina round |
| Marg. tex. | Marginulina texana |
| margas | marine gasoline |
| Mark | Markham |
| Marm | Marmaton |
| MARSH | Marshal (ling) |
| mass | massive |
| Mass pr. | Massilina pratti |
| Mat | matter |
| Math | mathematics |
| matl | material |
| MAW | mud acid wash |
| MAWP | maximum allowable working pressure |
| max | maximum |
| May | Maywood |
| MB | methylene blue |
| MB | Moody's Branch |
| MBF/D | thousand barrels fluid per day |
| Mbl Fls | Marble Falls |
| Mbo/d | thousand barrels oil per day |
| mbr | member (geologic) |
| MBTU | thousand British thermal units |
| MBW/D | thousand barrels of water per day |
| mc | megacycle |
| MC | mud cake |
| MC | mud cut |
| MC ls | Moore County lime |
| MCA | mud cleanout agent |
| MCA | mud-cut acid |
| MCB | master circuit board |
| McC | McClosky lime |
| MCC | motor control center |
| McCul | McCullough |
| McEl | McElroy |

| | |
|---|---|
| MCF | thousand cubic feet |
| Mcfd | thousand cubic feet per day |
| Mcfgpd | thousand cubic feet of gas per day |
| MCG | mud-cut gas |
| mchsm | mechanism |
| McK | McKee |
| McL | McLish |
| McMill | McMillan |
| MCO | mud-cut oil |
| MCP | maximum casing pressure |
| mcr-x | microcrystalline |
| MCSW | mud-cut salt water |
| MCT | computer-processed interpretation |
| MCW | mud-cut water |
| MD | measured depth |
| md | millidarcies |
| MD | Mt. Diablo |
| MDC | Monel drill collars |
| MDDO | maximum daily delivery obligation |
| MDF | market demand factor |
| mdl | middle |
| mdse | merchandise |
| MDQ | minimum daily quantity |
| Mdy | muddy |
| MEA | monoethanolamine |
| Meak | Meakin |
| meas | measure (ed) (ment) |
| mech | mechanic (al), mechanism |
| Mech DT | mechanical down time |
| med | median |
| Med | Medina |
| med | medium |
| Med B | Medicine Bow |
| med FO | medium fuel oil |

| | |
|---|---|
| med gr | medium grained |
| Medr | Medrano |
| Meet | Meeteetse |
| MEG | methane-rich gas |
| MEK | methylethylketone |
| memo | memorandum |
| Men | Menard lime |
| Mene | Menefee |
| MEOH | methanol |
| MEP | mean effective pressure |
| MER | maximum efficient rate |
| Mer | Meramec |
| merc | mercury |
| mercap | mercaptan |
| merid | meridian |
| Meso | Mesozoic |
| meta | metamorphic |
| meth | methane |
| meth-bl | methylene blue |
| meth-cl | methyl chloride |
| methol | methanol |
| methr | methanator |
| metr | metric |
| mev | million electron volts |
| mezz | mezzanine |
| MF | manifold |
| MF | mud filtrate |
| MF- | Miscellaneous Field Studies Map |
| MFA | male to female angle |
| mfd | manufactured |
| MFD | mechanical flow diagram |
| mfd | microfarad |
| mfg | manufacturing |
| MFP | maximum flowing pressure |

| | |
|---|---|
| MFR | manufacture (er) |
| mg | medium grained |
| mg | milligram |
| mg | motor generator |
| MG | multigrade |
| MG | thousand gallons |
| m'gmt | management |
| mgr | manager |
| MGS | middle ground shoals |
| MH | manhole |
| mh | millihenry |
| MH | mousehole |
| mho/m | mhos per meter |
| MHz | megahertz (megacycles per second) |
| MI | malleable iron |
| MI | mile (s) |
| MI | mineral interest |
| MI | moving in (equipment) |
| mica | mica, micaceous |
| MICR | moving in completion rig |
| micro-xin | microcrystalline |
| microsec | microsecond |
| MICT | moving in cable tools |
| MICU | moving in completion unit |
| mid | middle |
| Mid | Midway |
| MIDDU | moving (moved) in double drum unit |
| MIE | moving in equipment |
| MIK | methylisobutylketone |
| mil | military |
| mill | milliotitic |
| millg | milling |
| MIM | moving in materials |
| min | minerals |

| | |
|---|---|
| min | minimum |
| min | minute (s) |
| min P | minimum pressure |
| Minl | Minnelusa |
| Mio | Miocene |
| MIPU | moving in pulling unit |
| MIR | moving in rig |
| MIRT | moving in rotary tools |
| MIRU | moving in and rigging up |
| MIRUSU | moving in rigging up swabbing units |
| misc | miscellaneous |
| Mise | Misener |
| MISR | moving in service rig |
| Miss | Mississippian |
| Miss Cany | Mission Canyon |
| MIST | moving in standard tools |
| MIT | moving in tools |
| MIU | moisture impurities and unsaponifiabales (grease-testing) |
| mix | mixer |
| MIXG | mixing |
| MKG | making |
| mkt | market (ing) |
| Mkta | Minnekahta |
| mky | milky |
| ml | milliliter |
| ML | mud logger |
| ml TEL | milliliters tetraethyl lead per gallon |
| ml | milled |
| Mle | milled one end |
| mlg | milling |
| MLL | master load list |
| MLU | mud logging unit |
| MLW | mean low wave |

**Standard Oil & Gas Abbreviator**

| | |
|---|---|
| MLW-PLAT | mean low water to platform |
| Mly | marly |
| mm | millimeter |
| MM | million (i.e., 9MM = 9,000,000) |
| MM | motor medium |
| mm Hg | millimeters of mercury |
| MMBLS | millions of barrels |
| MMBTU | million British thermal units |
| MMcf | million cubic feet |
| MMcfd | million cubic feet per day |
| mmf | magnetomotive force |
| MMRVB | million reservoir barrels |
| MMS | Minerals Management Service |
| MMscfd | million standard cubic feet per day |
| MNL | manual |
| MNR | minor |
| mnrl | mineral |
| MO | molybdenum |
| MO | motor oil |
| MO | moving out |
| mob | mobile |
| MOCT | moving out (off) cable tools |
| MOCU | moving out completion unit |
| mod | model |
| mod | moderate (ly) |
| mod | modification |
| modu | modular |
| MOE | milled other end |
| MOE | moving out equipment |
| Moen | Moenkopi |
| mol | molas |
| mol | mole |
| MOL | molecule, molecular |
| mol | mollusca |

## Abbreviations with Definitions

| | |
|---|---|
| mol wt | molecular weight |
| mon | monitor |
| MON | motor octane number |
| Mont | Montoya |
| Moor | Mooringsport |
| MOP | maximum operating pressure |
| Mor | Morrow |
| MOR | moving out rig |
| Morr | Morrison |
| MORT | moving out (off) rotary tools |
| Mos | Mosby |
| mot | motor |
| mott | mottled |
| MOU | motor oil units |
| mov | moving |
| Mow | Mowry |
| MP | maximum pressure |
| MP | mechanical properties |
| MP | melting point |
| MP | multipurpose |
| mPa | megapascal |
| MPB | metal petal basket |
| MPGH-lith | multipurpose grease lithium base |
| MPGR-soap | multipurpose grease soap base |
| MPH | miles per hour |
| MPT | male pipe thread |
| MPY | miles per year |
| MR | marine rig |
| MR | meter run |
| Mr | milliroentgen |
| MRF | "merf"/main reaction furnace |
| MRK | marking |
| mrlst | marlstone |
| MRQ | memo requesting quote (s) |

| | |
|---|---|
| Ms | millisecond (s) |
| MS | motor severe |
| MSA | multiple service acid |
| MSC | mapping subcommittee |
| Mscf | thousand standard cubic feet |
| Mscf/d | thousand standard cubic feet per day |
| Mscf/h | thousand standard cubic feet per hour |
| MSDS | material safety data sheets |
| Msl | mean sea level |
| MSP | maximum surface pressure |
| MSR | mud/silt remover |
| Mstr | master |
| MSW | muddy salt water |
| MSWG | miscible substance group |
| MT | empty container |
| MT | magnetic particle examination |
| MT | marcaroni tubing |
| Mt. Selm | Mount Selman |
| MTD | maximum total depth |
| MTD | mean temperature difference |
| MTD | measured total depth |
| mtd | mounted |
| mtg | mounting |
| mtge | mortgage |
| mtl | material |
| MTO | material take-off |
| MTP | maximum top pressure |
| MTP | maximum tubing pressure |
| mtr | meter |
| MTR | motor |
| MTS | mud to surface |
| Mtx | matrix |
| mud wt | mud weight |
| mudst | mudstone |

## Abbreviations with Definitions

| | |
|---|---|
| MULTX | multiply, multiplexer |
| musc | muscovite |
| MUX | middle hydrocrackate |
| mv | millivolt |
| Mvde | Mesa Verde |
| MVFT | motor vehicle fuel tax |
| MVOP | monthly volume operation plan |
| MW | megawatt |
| MW | microwave |
| MW | mud weight |
| MW | muddy water |
| MWD | marine wholesale distributors |
| MWP | maximum working pressure |
| MWPE | mill wrapped plain end |
| Mwy | Midway |
| mxd | mixed |
| M1E | milled one end |
| M2E | milled two ends |
| m³/d | cubic meters per day |

| | |
|---|---|
| N | Newton |
| N | nonproducer |
| N | normal (to express concentration) |
| N | north |
| N. Cock. | Nonionella Cockfieldensis |
| N/O | north offset |
| N/S S/S | nonstandard service station |
| N/tst | no test |
| N/2 | north half |
| N/4 | north quarter |
| NA | not applicable |

| | |
|---|---|
| NA | not available |
| Nac | Nacotoch |
| nac | nacreous |
| NaCL | sodium chloride |
| NaCO$_3$ | sodium carbonate |
| NAG | no appreciable gas |
| NALRD | Northern Alberta Land Registration district |
| NaOH | sodium hydroxide |
| nap | naptha |
| NARR | narrative |
| nat | natural |
| nat'l | national |
| Nav | Navajo |
| Navr | Navarro |
| NB | new bit |
| NB | nitrogen blanket |
| Nbg | Newburg |
| NC | no change |
| NC | no core |
| NC | normally closed |
| NC | not completed |
| NCT | national coarse thread |
| NCT | noncontiguous tract |
| ND | nippled down |
| ND | nondetergent |
| ND | not drilling |
| NDBOPs | nipple (ed) (ing) down blowout preventers |
| NDE | not deep enough |
| NSG | no show gas |
| Ndl Cr | Noodle Creek |
| NDT | nipple-down tree |
| NDT | nondestructive testing |
| NE | nonemulsifying agent |
| NE | northeast |

## Abbreviations with Definitions

| | |
|---|---|
| NE/4 | northeast quarter |
| NEA | nonemulsion acid |
| NEC | National Electric Code |
| NEC | northeast corner |
| neg | negative |
| neg | negligible |
| NEGO | negotiation |
| NEL | northeast line |
| NEP | net effective pay |
| neut | neutral neutralization |
| Neut. No. | Neutralization Number |
| New Alb | New Albany shale |
| Newc | Newcastle |
| NF | National Fine (thread) |
| NF | natural flow |
| NF | no fluid |
| NF | no fluorescence |
| NF | no fuel |
| NFD | new field discovery |
| NFOC | no fluorescence or cut |
| NFW | new field wildcat |
| NG | natural gas |
| NG | no gauge |
| NG | no good |
| NGL | natural gas liquids |
| NGTS | no gas to surface |
| NHDS | naptha-hydrogen desulfurization |
| $NH_3$ | ammonia |
| $NH_4Cl$ | ammonium chloride |
| NIC | not in contract |
| NID | notice of intention to drill |
| Nig | Niagra |
| Nine | Ninnescah |
| Niob | Niobrara |

| | |
|---|---|
| nip | nipple |
| nitro | nitoglycerine |
| NL | north line |
| NL Gas | nonleaded gas |
| NLL | neutron lifetime log |
| N'ly | northerly |
| NMI | nautical line |
| NO | new oil |
| NO | Noble-Olson |
| NO | normally open |
| NO | number |
| No Inc | no increase |
| no rec | no recovery |
| No. | number (before a number, i.e., No. 3) |
| NOB | not on bottom |
| nod | nodule, nodular |
| Nod. blan. | Nodosaria blanpiedi |
| Nod. Mex. | Nodosaria mexicana |
| NOJV | nonoperated joint ventures |
| nom | nominal |
| Non | Nonionella |
| nonf G | nonflammable compressed gas |
| NOP | nonoperating property |
| NOR | no order required |
| nor | normal |
| NOV | notice of violation |
| noz | nozzle |
| NP | nameplate |
| NP | nickel plated |
| NP | no production |
| NP | nonporous |
| NP | not prorated |
| NP | not pumping |
| NP | notary public |

## Abbreviations with Definitions

| | |
|---|---|
| NPD | new pool discovery |
| NPDES | National pollution discharge elimination system |
| NPL | nipple |
| npne | neoprene |
| NPOS | no paint on seams |
| NPR | Naval Petroleum Reserve |
| NPRA | Naval Petroleum Reserve, Alaska |
| NPS | nominal pipe size |
| NPSH | net positive suction head |
| NPT | National pipe thread |
| NPTF | National pipe thread, female |
| NPTM | National pipe thread, male |
| NPW | new pool wildcat |
| NPX | new pool exempt (nonprorated) |
| NR | new rod |
| NR | no recovery |
| NR | no report, not reported |
| NR | nonreturnable, no returns, not reached |
| NRI | net revenue interest |
| NRS | nonrising stem (valve) |
| NRSB | nonreturnable steel barrel |
| NRSD | nonreturnable steel drum |
| NS | no show |
| NSC | not suitable for coating |
| NSFOC | no show fluorescence or cut |
| NSO | no show oil |
| NOS&G | no show oil and gas |
| NSPS | new source performance standards |
| nstd | nonstandard |
| NT | net tons |
| NT | no time |
| NTD | new total depth |
| NTP | notice to proceed |
| NTS | not to scale |

| | |
|---|---|
| NU | naphfining unit |
| NU | nippled (ing) up |
| NU | nonupset |
| NUBOPs | nipple (ed) (ing) up blowout prevents |
| NUE | nonupset ends |
| Nug | nugget |
| num | numerous |
| NUT | nipple up tree |
| NUWH | nippling up wellhead |
| NVP | no visible porosity |
| NW | no water |
| NW | northwest |
| NW/C | northwest corner |
| NW/4 | northwest quarter |
| NWL | northwest line |
| NWT | Northwest Territories |
| NYA | not yet available |
| NYD | not yet drilled |
| NYL | nylon |
| $N_2$ | nitrogen |

## O

| | |
|---|---|
| O | oil |
| O | Osborne |
| O sd | oil sand |
| O&G | oil and gas |
| O&GC SULW | oil and gas-cut sulfur water |
| O&GCAW | oil and gas-cut acid water |
| O&GCLW | oil and gas-cut load water |
| O&GCM | oil and gas-cut mud |
| O&GCSW | oil and gas-cut salt water |
| O&GCW | oil and gas-cut water |

## Abbreviations with Definitions

| | |
|---|---|
| O&GL | oil and gas lease |
| O&M | operations and maintenance |
| O&SW | oil and salt water |
| O&SWCM | oil and sulfur water-cut mud |
| O&W | oil and water |
| O/S | out of service over and short (report) |
| O/S | out of stock |
| O/T tk | open-top tank |
| OA | overall |
| OAH | overall height |
| Oakv | Oakville |
| OAL | overall length |
| OAW | oil abandoned well |
| OB | off bottom |
| obj | object |
| OBM | oil-based mud |
| OBMO | outboard motor oil |
| OBS | observation |
| OBS | ocean bottom suspension |
| obsol | obsolete |
| OBW&RS | optimum bit weight and rotary speed |
| OC | oil cut |
| OC | on center |
| OC | open choke |
| OC | open cup |
| OC | operations commenced |
| OC- | Oil and Gas Investigations Chart |
| OCB | oil circuit breaker |
| occ | occasional (ly) |
| OCM | oil-cut mud |
| OCS | Outer Continental Shelf |
| OCSW | oil-cut salt water |
| oct | octagon, octagonal |
| oct | octane |

| | |
|---|---|
| OCW | oil-cut water |
| od | odor |
| OD | outside diameter |
| Odel | O'Dell |
| ODT | oil down to |
| OE | oil emulsion |
| OE | open end |
| OE | overexpenditure |
| OEB | other end beveled |
| OEM | oil emulsion mud |
| OF | open flow |
| OF | open-file report |
| off | office, official |
| off-sh | offshore |
| OFIC | oil insulated fan-cooled |
| OFL | overflush (ed) |
| OFLU | oil fluorescence |
| OFOE | orfice flange one end |
| OFP | open flow potential |
| OFS | offsite |
| O'H | O'Hara |
| OH | open hearth |
| OH | open hole |
| OH | overhead |
| ohm | ohm |
| ohm-cm | ohm-centimeter |
| ohm-m | ohmmeter |
| OI | oil insulated |
| OIFC | oil insulated, fan cooled |
| OIH | oil in hole |
| Oil Cr | Oil Creek |
| oilfract | oil fractured |
| OIP | oil in place |
| OISC | oil insulated, self-cooled |

## Abbreviations with Definitions

| | |
|---|---|
| OIT | oil in tanks |
| OIWC | oil immersed, water cooled |
| OL | off / on location |
| OL | open line (no choke) |
| ole | olefin |
| Olig | Oligocene |
| OLN | outline |
| OMRL | oriented microresistivity |
| ONR | octane number requirement |
| ONRI | octane number requirement increase |
| OO | oil odor |
| ooc | ooliclastic |
| OOIP | original oil in place |
| ool | oolitic |
| oom | oolimoldic |
| OP | articles published in outside journals/books |
| OP | oil pay |
| OP | outpost |
| OP | overproduced |
| op hole | open hole |
| OPB | old plugback |
| OPBD | old plugback depth |
| oper | operate, operations, operator |
| Operc | Operculinoides |
| OPI | oil payment interest |
| opn | open (ed) (ing) |
| OPO | overseas procurement office |
| opp | opposite |
| OPT | official potential test |
| OPTL | optional |
| optn to F/O | option to farmout |
| OR | orange |
| Or | Oread |
| Ord | Ordovican |

| | |
|---|---|
| orf | orifice |
| org | organic |
| org | organization |
| ORIENT | orientation |
| orig | original, originally |
| Orisk | Oriskany |
| ORR | overriding royalty |
| ORRI | overriding royalty interest |
| orth | orthoclase |
| OS | oil show |
| Os | Osage |
| OS | overshot |
| OS&F | odor stain and fluorescence |
| OS&Y | outside screw and yoke (valve) |
| OSA | oil-soluble acid |
| OSD | operation shutdown |
| OSF | oil string flange |
| OSI | oil well shut in |
| OSIDP | oil standing in drillpipe |
| Ost | Ostracod |
| OSTN | oil stain |
| OSTOIP | original stock tank oil in place |
| Osw | Oswego |
| OT | open tubing |
| OT | overtime |
| OT&S | odor taste & stain |
| OTD | old total depth |
| OTD | original total depth |
| OTE | oil-powered total energy |
| Otl | outlet |
| OTS | oil to surface |
| OTS&F | odor taste stain and fluorescence |
| OU | oil unit |
| Our | Ouray |

## Abbreviations with Definitions

| | |
|---|---|
| Ovhd | overhead |
| OWC | oil-water contact |
| OWDD | oil well drilled deeper |
| OWF | oil well flowing |
| OWFWF | oil well from water flood |
| OWG | oil well gas |
| OWPB | old well plugged back |
| OWST | old well sidetracked |
| OWWO | old well worked over |
| ox | oxidized, oxidation |
| oxy | oxygen |
| oz | ounce |

| | |
|---|---|
| P | professional paper |
| P&A | plugged and abandoned |
| P&ID | process & instrument diagram |
| P&NG | petroleum and natural gas |
| P Lar | Post Laramine |
| P tstg | pump testing |
| p. | page (before a number, i.e., p. 4) |
| P.O. | Pin Oak |
| P.O. | Post Oak |
| P.P. | present production |
| P-HDII | perforating Hyperdome II |
| P-M | Pensky-Martins (flash) |
| P&C | personal and confidential |
| P&F | pump and flow |
| P&IDS | piping and instrument diagrams |
| P&L | profit and loss |
| P&P | porosity and permeability |
| P&P | porous and permeable |

| | |
|---|---|
| P/ | pump |
| P/BLDG | pump building |
| P/DIA | piping diagram |
| PA | participating area |
| Pa | Pascal |
| PA | pooling agreement |
| PA | pressure alarm |
| PA | public address |
| PAB | per-acre bonus |
| Padd | Paddock |
| Paha | Pahasapa |
| Pal | Paluxy |
| Paleo | paleontology |
| Paleo | Paleozoic |
| Palo P | Palo Pinto |
| Pan L | Panhandle Lime |
| PAR | per-acre rental |
| Para | Paradox |
| Park C | Park City |
| PART | partial |
| pat | patent (ed) |
| patn | pattern |
| pav | paving |
| Paw | Pawhuska |
| PB | plugged back |
| PB | pressure base |
| PB-ADA | report available only through National Technical Information Service |
| PBD | plugged-back depth |
| PBE | plain both ends |
| PBHL | proposed bottom-hole location |
| pbl (y) | pebble, pebbly |
| PBP | pulled bid pipe |
| PBTD | plugged back total depth |

## Abbreviations with Definitions

| | |
|---|---|
| PBW | pipe buttweld |
| PBX | private branch exchange |
| PBX | switchboard |
| PC | Paint Creek |
| pc | piece |
| PC | poker chipped |
| pc | port collar |
| PC | Porter Creek |
| PCF | pounds per cubic foot |
| pct | percent |
| PCV | positive crankcase ventilation |
| PCV | pressure control valve |
| PD | geopressure development, failure |
| pd | paid |
| PD | per day |
| PD | plug down |
| PD | present depth |
| PD | pressed distillate |
| PD | proposed depth |
| PD | pulsation dampner |
| PD | pumper's depth |
| PDC | power distribution center |
| PDC | pressure differential controller |
| PDET | production department exploratory test |
| PDI | pressure differential indicator |
| PDIC | pressure differential indicator controller |
| PDR | pressure differential recorder |
| PDRC | pressure differential recorder controller |
| PDS | geopressure development, success |
| PDS | power distribution system |
| pdso | pseudo |
| pe | pin end |
| PE | plain end |
| PE | pumping equipment |

| | |
|---|---|
| PEB | plain end beveled |
| PED | pedestal |
| pell | pellletal, pelletoidal |
| pen | penetration, penetration test |
| Pen A.C. | penetration asphalt cement |
| penal | penalty, penalize (ed) (ing) |
| Penn | Pennsylvanian |
| perco | percolation |
| perf | perforate (ed) (ing) (or) |
| perf csg | perforated casing |
| perm | permanent |
| perm | permeable (ability) |
| Perm | Permian |
| perp | perpendicular |
| pers | personnel |
| PERT | performance evaluation and review technique |
| pet | petroleum |
| Pet | Pettet |
| Pet sd | Pettus sand |
| petrf | petroliferous |
| petrochem | petrochemical |
| Pett | Pettit |
| PEW | pipe electric weld |
| pf | per foot |
| PF | power factor |
| pfd | preferred |
| PFD | process flow diagram |
| PFM | power factor meter |
| PFRACT | prefractionator |
| PFT | pumping for test |
| PG | Pecan Gap |
| Pg | plug |
| PGC | Pecan Gap chalk |
| PGW | producing gas well |

## Abbreviations with Definitions

| | |
|---|---|
| pH | acidity or alkalinity |
| pH | hydrogen ion concentration |
| pH | measure of hydrogen potential |
| Ph | parish |
| ph | phase |
| PHC | pipe-handling capacity |
| Phos | Phosphoria |
| PI | penetration index |
| PI | Pine Island |
| PI | pressure indicator |
| PI | productivity index |
| PI | pump in |
| PIC | pressure indicator controller |
| Pic Cl | Pictured Cliff |
| pinpt | pinpoint |
| PIP | pump-in pressure |
| piso | pisolites, pisolitic |
| pit | pitted |
| PJ | pump jack |
| PJ | pump job |
| pk | pink |
| pkg (d) | packing, package (ed) |
| pkr | packer |
| PL | pipeline |
| PL | plate |
| PL | property line |
| pl fos | plant fossils |
| Plan. palm. | Planulina palmarie |
| Plan. har. | Planulina harangensis |
| plas | plastic |
| PLASR | plaster |
| platf | platform |
| plcy | pelecypod |
| pld | pulled |

| | |
|---|---|
| PLE | plain large end |
| Pleist | Pleistocene |
| plg | plagioclase |
| plg | pulling |
| plgd | plugged |
| Plio | Pliocene |
| PLMB | plumbing |
| pln | plan |
| plngr | plunger |
| PLO | pipeline oil |
| PLO | pumping load oil pilot |
| plt | pilot |
| PLT | pipeline terminal |
| plt | plant |
| plty | platy |
| PLV | pilot loaded valve |
| PLW | pipe lapweld |
| P-M | Pensky Martins |
| PML | production management |
| pmp (d) (g) | pump (ed) (ing) |
| pmt | payment |
| PN | Performance Number (aviation gas) |
| pneu | pneumatic |
| pnl | panel |
| PNL BD | panel board |
| PNR | please note and return |
| po | Phrohotite |
| PO | pulled out |
| PO | pumps off |
| PO | purchase order |
| POB | plug on bottom |
| POB | pump on beam |
| POCS | Pacific Outer Continental Shelf |
| POD | plan of development |

| | |
|---|---|
| Pod. | Podbielniak |
| POE | plain one end |
| POGW | producing oil and gas well |
| POH | pulled (put) out of hole |
| pois | poison |
| pol | polish (ed) |
| poly | polymerization, polymerized |
| poly cl | polyethylene |
| polygas | polymerized gasoline |
| polypl | polypropylene |
| PONA | parrafins-olefins-napthenes-aromatics |
| Pont | Pontotoc |
| POOH | pull (put) out of hole |
| POP | putting on pump |
| por | porosity, porous |
| porc | porcelaneous |
| porc | porcion |
| port | portable |
| pos | position |
| pos | positive |
| poss | possible (ly) |
| pot | potential |
| pot dif | potential difference |
| POT/WTR | potable water |
| pour ASTM | pour point (ASTM method) |
| POW | producing oil well |
| POWF | producing oil well, flowing |
| POWP | producing oil well, pumping |
| PP | pinpoint |
| PP | production payment |
| PP | pulled pipe |
| PP | pump pressure |
| ppb | parts per billion |
| PPB | pounds per barrel |

| | |
|---|---|
| ppd | prepaid |
| ppg | piping |
| PPG | pounds per gallon |
| PPI | production payment interest |
| PPI | Process Performance Index |
| ppm | parts per million |
| ppn no | precipitation number |
| PPP | pinpoint porosity |
| ppt | precipitate |
| pr | pair |
| PR | polished rod |
| pr | poor |
| PR | pressure recorder |
| PR | public relations |
| PR | purchasing request |
| pr op | present operations |
| PR&T | pull (ed) rods and tubing |
| PRC | pressure recorder control |
| prcst | precast |
| prd | period |
| Pre Camb | Precambrian |
| PRECIP | precipitator |
| predom | predominant |
| prefab | prefabricated |
| prehtr | preheater |
| prelim | preliminary |
| prem | premium |
| Prep | prepare, preparing, preparation |
| press | pressure |
| prest | prestressed |
| prev | prevent, preventive |
| PREV | previous |
| PREV DA AVG | previous daily average |
| PRF | Primary Reference Fuel |

| | |
|---|---|
| pri | primary |
| prin | principal |
| priv | privilege |
| prly | pearly |
| prmt | permit |
| prncpl lss | principal lessee (s) |
| pro | prorated |
| prob | probable (ly) |
| proc | process |
| prod | produce (ed) (ing) (tion), product (s) |
| prog | progress |
| proj | project (ed) (ion) |
| PROP | property |
| prop | proportional |
| prop | propose (ed) |
| prot | protection |
| Protero | Proterozoic |
| Prov | provincial |
| prsm | prism(atic) |
| PRPT | preparing to take potential test |
| prtgs | partings |
| PS | pressure switch |
| ps | pseudo |
| PS | pump station |
| PSA | packer set at |
| PSB | pressure seal bonnet |
| PSD | permanently shut down |
| PSD | prevention of significant deterioration, EPA |
| PSE | plain small end |
| psf | pounds per square foot |
| psi, PSI | pounds per square inch |
| PSI | profit-sharing interest |
| psia, PSIA | pounds per square inch absolute |
| psig, PSIG | pounds per square inch gauge |

| | |
|---|---|
| PSL | pipe sleeve |
| PSL | Public School Land |
| PSM | pipe seamless |
| Psp | prospect |
| PSV | pressure safety valve |
| PSW | pipe spiral weld |
| PT | liquid penetrant examination |
| pt | party, partly |
| pt | pint |
| pt | point |
| Pt Lkt | Point Lookout |
| PT | potential test |
| PTC | permanent type completion |
| PTD | present total depth |
| PTD | projected total depth |
| PTD | proposed total depth |
| PTF | production test flowed |
| PTG | pulling tubing |
| PTN | partition |
| PTP | production test pumped |
| PTR | pulling tubing and rods |
| PTS pot | pipe to soil potential |
| PTTF | potential test to follow |
| PU | picked up |
| PU | pulled up |
| PU | pumping unit |
| PUC | project ultimate cost |
| PUDP | picking up drillpipe |
| PUIC | pulled up in casing |
| PULS | pulse (sating) (sation) |
| PURCH | purchasing |
| PURF | purification |
| purp | purple |
| PV | plastic viscosity |

## Abbreviations with Definitions

| | |
|---|---|
| PV | pore volume |
| PVC | polyvinyl chloride |
| pvmt | pavement |
| PVR | plant (pressure) volume reduction |
| PVT | pressure-volume temperature |
| PW | producing well |
| PW(15) | present worth at discount rate of 15% |
| PWHT | postweld heat treatment |
| PWR | power |
| PWY | pipeway |
| PWZ | peripheral wedge zone |
| Pxy | Paluxy |
| pyls | pyrolysis |
| pyr | pyrite, pyritic |
| pyrbit | pyrobitumen |
| pyrclas | pyroclastic |

## Q

| | |
|---|---|
| Q. City | Queen City |
| Q. sd | Queen Sand |
| QA | quality assurance |
| QC | quality control |
| QDA | quality discount allowance |
| QDRNT | quadrant |
| qnch | quench |
| QRC | quick ram change |
| qry | quarry |
| qt | quart (s) |
| qtr | quarter |
| qty | quantity |
| qtz | quartz, quartzite, quartzitic |
| qtzose | quartzose |

Standard Oil & Gas Abbreviator

| | |
|---|---|
| quad | quadrant, quadrangle, quadruple |
| QUAL | qualitative |
| qual | quality |
| quan | quantity |
| quest | questionable |
| quint | quintuplicate |

| | |
|---|---|
| R | radius |
| R | range |
| R | rankine (temp.scale) |
| R | resistivity |
| r | roentgen |
| R | rows |
| R&D | research and development |
| R&T | rods and tubing |
| R test | rotary test |
| R.O. | Red Oak |
| R(16") | resistivity (as recorded from 16" electrode configuration) |
| R-SP | recommended spare part |
| R&L | road & location |
| R&LC | road & location complete |
| R&O | rust and oxidation |
| R/A | regular acid |
| RA | radioactive |
| RA | right angle |
| RACTR | reactor |
| rad | radical |
| rad | radiological |
| rad | radius |
| RADT | radiant |

| | |
|---|---|
| radtn | radiation |
| RAGL | raw gas lift |
| RALOG | running radioactive log |
| Rang | Ranger |
| RB | rock bit |
| RB | rotary bushing |
| RBLR | reboiler |
| Rbls | rubber balls |
| RBM | rotary bushing measurement |
| RBP | retrievable bridge plug |
| rbr | rubber |
| RBSO | rainbow show of oil |
| RBSOF | rubber ball sand oil frac |
| RBSWF | rubber ball sand water frac |
| RC | rapid curing |
| RC | Red Cave |
| RC | remote control |
| RC | reverse circulation |
| RC | running casing |
| RCO | returning circulation oil |
| RCPT | receptacle |
| RCR | Ramsbottom Carbon Residue |
| RCR | reverse circulation rig |
| RCTN | reaction |
| RCVR | receiver |
| RCVY | recovery |
| RCYL | recycle |
| RD | redrilled |
| RD | rigged (ing) down |
| rd | road |
| rd | round |
| Rd Bds | red beds |
| Rd Fk | Red Fork |
| Rd Pk | Red Peak |

| | |
|---|---|
| rd thd | round thread |
| RDB | rotary drive bushing |
| RDB-GD | rotary drive bushing to ground |
| RDCR | reducer |
| rdd | rounded |
| RDMO | rigged down, moved out |
| RDS | reservoir description service |
| RDSU | rigged-down swabbing unit |
| rdtp | round trip |
| REABS | reabsorber |
| reacd | reacidize (ed) (ing) |
| React | reaction (ed) |
| rebar | reinforcing bar |
| rec | recommend |
| rec | record (er) (ing) |
| rec | recover (ed) (ing), recovery |
| recd | received |
| recip | reciprocate (ing) |
| recirc | recirculate |
| recomp | recomplete (ed) (ion) |
| RECOMP | recompressor |
| recond | recondition (ed) |
| recp | receptacle |
| rect | rectangle, rectangular |
| rect | rectifier |
| recy | recycle |
| red | reducing, reducer |
| red bal | reducing balance |
| redrld | redrilled |
| ref | reference |
| ref | refine (ed) (er) (ry) |
| refg | refining |
| refl | reflection |
| refl | reflux |

| | |
|---|---|
| REFMR | reformer |
| REFOO | re-evaluation for overoptimism |
| reform | reformate (er) (ing) |
| refr | refraction, refractory |
| REFRIG | refrigerator (rant) (tion) |
| reg | register |
| reg | regular, regulator |
| regen | regenerator |
| reinf | reinforce (ed) (ing) (ment) |
| reinf conc | reinforced concrete |
| rej | reject |
| rej'n | rejection |
| Rek | Reklaw |
| rel | relay |
| rel | release (ed) |
| REL | running electric log |
| reloc | relocate (ed) |
| rem | remains |
| rem | remedial |
| Ren | Renault |
| rent | rental |
| Reo. bath. | Reophax bathysiphoni |
| rep | repair (ed) (ing) (s) |
| rep | replace (ed) |
| rep | report |
| reperf | reperforated |
| repl | replace (ment) |
| REQ | request |
| req | requisition |
| reqd | required |
| reqmt | requirement |
| res | research |
| res | reserve (ation) |
| res | resistance, resistivity, resistor |

| | |
|---|---|
| Res. O. N. | Research Octane Number |
| resid | residual, residue |
| RESIS | resistor (s) |
| ret | retain (er) (ed) (ing) |
| ret | return |
| retd | returned |
| retr | retrievable |
| retr ret | retrievable retainer |
| rev | reverse (ed) |
| rev | revise (ed) (ing) (ion) |
| rev | revolution (s) |
| rev/O | reversed out |
| RF | raised face |
| RF | rig floor |
| RFFE | raised face, flanged end |
| RFG | roofing |
| RFG/BD | refrigeration building |
| RFLCT | reflect (ed) (ing) (tion) |
| RFP | request for proposal |
| RFQ | request for quote |
| RFR | ready for rig |
| RFSF | raised face, smooth finish |
| RFSO | raised face, slip on |
| RFWN | raised face, weld neck |
| RG | raw gas |
| rg | ring |
| RG | ring groove |
| Rge | range |
| rgh | rough |
| RGTR | register |
| RH | rat hole |
| RH | relative humidity |
| RH | right hand |
| RHD | righthand door |

## Abbreviations with Definitions

| | |
|---|---|
| rheo | rheostat |
| RHM | rat hole mud |
| RHN | Rockwell hardness number |
| RI | royalty interest |
| Rib | ribbon sand |
| Rier | Rierdon rig |
| RIH | ran in hole |
| RIL | red indicating lamp |
| riv | rivet |
| RIZ | resistivity invaded zone |
| RJ | ring joint |
| RJFE | ring joint, flanged end |
| RK | rack |
| rk | rock |
| RKB | rotary kelly bushing |
| rky | rocky |
| RL | random lengths |
| rlf | relief |
| rlg | railing |
| rls (d) (ing) | release (ed) (ing) |
| rly | relay |
| rm | ream |
| Rm | resistivity, mud |
| rm | room |
| rmd | reamed |
| Rmf | resistivity, mud filtrate |
| rmg | reaming |
| rmn | remains |
| RMS | root mean square |
| rmv (l) | remove (al) (able) |
| rnd | rounded |
| rng | running |
| RO | reversed out |
| ro | rose |

| | |
|---|---|
| Ro | Rosiclare sand |
| Rob | Robulus |
| Rod | Rodessa |
| ROF | rich oil fractionator |
| ROGL | rotative gas lift |
| ROI | return on investment |
| Rok | remove |
| Rok | rock |
| ROL | rig on location |
| ROM | rough order of magnitude |
| ROM | run of mine |
| RON | Research Octane Number |
| ROP | rate of penetration |
| ROR | rate of return |
| ROS | remote operating system (station) |
| rot | rotary, rotate, rotator |
| ROW | right of way |
| roy | royalty |
| RP | rock pressure |
| rpm | revolutions per minute |
| rpmn | repairman |
| RPP | retail pump price |
| RPRT | report |
| rps | revolutions per second |
| rptd | reported |
| RR | railroad |
| RR | Red River |
| RR | rig released |
| RR | rig repair |
| RR | rigging rotary |
| RR&T | ran (running) rods and tubing |
| RS | rig service |
| RS | rig skidded |
| RS | rising stem (valve) |

## Abbreviations with Definitions

| | |
|---|---|
| RSD | returnable steel drum |
| rsns | resinous |
| RSU | released swab unit |
| rsvr | reservoir |
| RT | radiographic examination |
| RT | rig time |
| RT | rotary time |
| RT | rotary tools |
| RT CB | round trip changed bit |
| rtd | retard (ed) |
| RTD | rotary total depth |
| rtg | rating |
| RTG | routing |
| RTG | running tubing |
| RTJ | ring tool joint |
| RTJ | ring-type joint |
| RTL | Refinery Technology Laboratory |
| RTLTM | rate too low to measure |
| RTMTR | rotameter |
| rtnr | retainer |
| RTTS | retrievable test treat squeeze (tool) |
| RTU | remote terminal unit |
| RU | rig (ged) (ging) up |
| RU | rotary unit |
| RUCC | rig-up casing crew |
| RUCT | rigging-up cable tools |
| RUM | rigging-up machines |
| RUP | rigging-up pump |
| rupt | rupture |
| RURT | rigging-up rotary tools |
| RUSR | rigging-up service rig |
| RUST | rigging-up standard tools |
| RUSU | rigging-up swabbing unit |
| RUT | rigging-up tools |

| | |
|---|---|
| RV | relief valve |
| RVP | Reid vapor pressure |
| rvs (d) | reverse (ed) |
| RVT | rivet |
| Rw | resistivity, water |
| Rwa | resistivity, water (apparent) |
| rwk (d) | rework (ed) |
| RWTP | returned well to production |
| Rxo | resistivity, flushed zone |

## S

| | |
|---|---|
| S | seconds |
| S | south |
| S | stratigraphic test |
| s&s | spigot and spigot |
| S&T | shell and tube |
| S Bomb | sulfur by bomb method |
| S O | south offset |
| S Riv | Seven Rivers |
| S.L. | sea level |
| S-T-R | section-township-range |
| S&F | swab and flow |
| S&O | stain and odor |
| s&p | salt and pepper |
| S/C | speed/current |
| S/E | screwed end |
| S/FAB | shop fabrication |
| s/p | shipping point (purchasing term) |
| S/SR | sliding-scale royalty |
| S/SW | screwed and socketweld |
| S/T | sample tops |
| S/T | speed/torque |

## Abbreviations with Definitions

| | |
|---|---|
| S/WTR | sanitary water |
| S/2 | south half |
| s,t&b | sides, tops, & bottoms |
| SA | seal assembly |
| Sab | Sabinetown |
| sach | saccharoidal |
| Sad Cr | Saddle Creek |
| sadl | saddle |
| saf | safety |
| SAF/DPT | safety/department |
| SAFE | surface approximation and formation evaluation |
| Sal | Salado |
| sal | salary, salaried |
| sal | salinity |
| Sal Bay | Saline Bayou |
| salv | salvage |
| samp | sample sanitary |
| SAN | sanitary |
| San And | San Andres |
| San Ang | San Angelo |
| San Raf | San Rafael |
| Sana | Sanastee |
| sani | sanitary |
| sap | saponification |
| Sap No. | saponification number |
| Sara | Saratoga |
| sat | saturated, saturation |
| Saw | Sawatch |
| Sawth | Sawtooth |
| Say Furol | Saybolt furol |
| SB | sideboom |
| SB | sleeve bearing |
| SB | stuffing box |
| sb | sub |

| | |
|---|---|
| Sb | Sunburst |
| SBA | secondary butyl alcohol |
| SBB&M | San Bernardino base and meridian |
| SBHP | static bottom-hole pressure |
| sc | scales |
| SC | self-contained |
| SC | show condensate |
| SC DL | slip and cut drill time |
| SCAF | scaffolding |
| scat (d) | scatter (ed) |
| scf, SCF | standard cubic foot |
| scfd, SCFD | standard cubic feet per day |
| scfh, SCFH | standard cubic feet per hour |
| scfm, SCFM | standard cubic feet per minute |
| sch | schedule |
| schem | schematic |
| scly | securaloy |
| scolc | scolescodonts |
| scr | scraper |
| scr | scratcher |
| scr | screen |
| scr (d) | screw (ed) |
| scrub | scrubber |
| SCSSV | surface-controlled subsurface safety valve |
| sctrd | scattered |
| sd, SD | sand |
| SD | shut down |
| sd&sh | sand and shale |
| SD Ck | side door choke |
| Sd SG | sand showing gas |
| Sd SO | sand showing oil |
| SDA | shut down to acidize |
| sdd | sanded |
| SDFN | shut down for night |

| | |
|---|---|
| SDF | shut down to fracture |
| sdfract | sandfracked |
| SDG | siding |
| SDL | shut down to log |
| SDO | show of dead oil |
| SDO | shut down for orders |
| sdoilfract | sand-oil fracked |
| SDON | shut down overnight |
| SDP | set drillpipe |
| SDPA | shut down to plug and abandon |
| SDPL | shut down for pipeline |
| SDR | shut down for repairs |
| sdtkr | sidetrack (ed) (ing) |
| SDW | shut down for weather |
| SDWL | sidewall |
| SDWO | shut down awaiting orders |
| sdwtrfract | sand-water fracked |
| sdy | sandy |
| sdy li | sandy lime |
| sdy sh | sandy shale |
| SE | southeast |
| SE NA | screw end American National Acme thread |
| SE NC | screw end American National Coarse thread |
| SE No. | steam emulsion number |
| SE NTP | screw end American National Taper Pipe thread |
| SE/C | southeast corner |
| SE/4 | southeast quarter |
| Sea | Seabreeze |
| sec | secant |
| sec | second (ary) |
| sec | secretary |
| sec | section |
| SECT | section (s) (al) (ing) |
| sed | sediment (s) |

| | |
|---|---|
| Sedw | Sedwick |
| SEG | segment |
| seis | seismograph, seismic |
| sel | selenite |
| Sel | Selma |
| SELECT | selection (tive) (tor) |
| Sen | Senora |
| SEO | seal oil |
| SEP | separate, separator, separation |
| sept | septuplicate |
| seq | sequence |
| ser | series, serial |
| Serp | serpentine |
| Serr | Serratt |
| serv | service (s) |
| serv chg | service charge |
| set | settling |
| sew | sewer |
| SEWOP | self-elevating work platform |
| Sex | sexton |
| sext | sextuplicate, sextuplet |
| SF | sandfrac |
| sfc | surface |
| SFD | system flow diagram |
| SFL | starting fluid level |
| SFLU | slight, weak, or poor fluorescence |
| SFO | show of free oil |
| SFP | surface flow pressure |
| SFT | sample formation tester |
| sft | soft |
| SG | show gas |
| SG | surface geology |
| SG&W | show gas and water |
| SG&C | show gas and condensate |

| | |
|---|---|
| SG&D | show gas and distillate |
| SG&O | show gas and oil |
| SGCM | slightly gas-cut mud |
| SGCO | slightly gas-cut oil |
| SGCSW | slightly gas-cut salt water |
| SGCW | slightly gas-cut water |
| SGCWB | slightly gas-cut water blanket |
| SGCWC | slightly gas-cut water cushion |
| sgd | signed |
| sgl (s) | single (s) |
| sh | shale |
| sh | sheet |
| SH | substructure height |
| Shan | Shannon |
| SHDP | slim-hole drillpipe |
| Shin | Shinarump |
| shld | shoulder |
| shls | shells |
| SHLT | shelter |
| shly | shaly |
| SHM | solar heat medium |
| shp | shaft horsepower |
| shp(g) | ship (ping) |
| shpt | shipment |
| shr | shear |
| SHT | straight-hole test |
| SHTG | sheeting |
| SHTG | shortage |
| shthg | sheathing |
| SI | shut in |
| SIBHP | shutin bottom-hole pressure |
| SICP | shutin casing pressure |
| SIGW | shutin gas well |
| SIH | started in hole |

| | |
|---|---|
| Sil | Silurian |
| silic | silica, siliceous |
| silt | siltstone |
| sim | similar |
| Simp | Simpson |
| SIOW | shutin oil well |
| SIP | shutin pressure |
| Siph. d. | Siphonina davisi |
| SITP | shutin tubing pressure |
| SIWHP | shutin wellhead pressure |
| SIWOP | shutin, waiting on potential |
| sk | sack (s) |
| SK | sketch |
| Sk Crk | Skull Creek |
| skim | skimmer |
| Skn | Skinner |
| skr d | sucker rod |
| sks | slickensided |
| skt | socket |
| SL | section line |
| sl | sleeve |
| SL | south line |
| SL | state lease |
| SLC | steel line correction |
| sld | sealed |
| sli | slight (ly) |
| Sli | Sligo |
| sli SO | slight show of oil |
| slky | silky |
| SLM | steel line measurement |
| SLNCR | silencer |
| slnd | solenoid |
| SLPR | sleeper |
| Slt Mtn | Salt Mountain |

## Abbreviations with Definitions

| | |
|---|---|
| Slty | salty |
| slty | silty |
| SLUR | slurry |
| slv | sleeve |
| S'ly | southerly |
| SM | Seward Meridian (Alaska) |
| sm | small |
| SM | surface measurement |
| Smithw | Smithwick |
| Smk | Smackover |
| smls | seamless |
| smpl | sample |
| smth | smooth |
| SN | seating nipple |
| SNG | synthetic natural gas |
| SNUB | snubber, snubbing |
| SNUFF | snuffing |
| SO | shake out |
| SO | shaled out |
| SO | show of oil side opening |
| SO | side opening |
| SO | slip on |
| SO&G | show oil and gas |
| SO&GCM | slightly oil- and gas-cut mud |
| SO&W | show oil and water |
| SOC | start of cycle |
| SOCM | slight oil-cut mud |
| SOCSW | slight oil-cut salt water |
| SOCW | slight oil-cut water |
| SOCWB | slight oil-cut water blanket |
| SOCWC | slight oil-cut water cushion |
| sod gr | sodium-based grease |
| SOE | screwed on one end |
| SOF | sand-oil fracture |

| | |
|---|---|
| SOH | shot open hole |
| SOH | started out of hole |
| sol | solenoid |
| sol | solids |
| soln | solution |
| solv | solvent |
| som | somastic |
| somct | somastic coated |
| SONL | sonic log |
| SOOH | started out of hole |
| SOOH | strapped out of hole |
| SOP | standard operational procedure |
| SOR | start of run |
| sort | sorted (ing) |
| SOV | solenoid-operated valves |
| sow | socket weld |
| SP | self (spontaneous) potential |
| SP | set plug |
| sp | shipping point (purchasing term) |
| sp | shot point |
| sp | slightly porous |
| sp | spare |
| Sp | Sparta |
| sp | spore |
| SP | straddle packer |
| SP | surface pressure |
| sp | shipping point (purchasing term) |
| sp gr | specific gravity |
| sp ht | specific heat |
| SP SW | single pole switch |
| sp. vol. | specific volume |
| SP-DST | straddle packer drillstem test |
| spcl | special |
| spcr | spacer |

# Abbreviations with Definitions

| | |
|---|---|
| spd | spud (ded) (der) |
| spdl | spindle |
| SPDT | single-pole double throw |
| SPDT SW | single-pole double throw switch |
| spec | specification |
| speck | speckled |
| SPF | shot per foot |
| spf | spearfish |
| spg | sponge |
| spg | sponge |
| SPGT | spigot |
| sph | spherules |
| Sphaer | Sphaerodina |
| sphal | sphalerite |
| spic | spicule (ar) |
| Spiro. b. | Spiroplectammina barrowi |
| spkr | sprinkler |
| spkt | sprocket |
| spl | sample |
| spl cham | sample chamber |
| spletp | Spindletop |
| SPLTR | splitter |
| splty | specialty |
| splty | splintery |
| SPM | strokes per minute |
| Spra | Spraberry |
| sprf | spirifers |
| Sprin | Springer |
| SPST | single-pole single throw |
| SPST SW | single-pole single throw stitch |
| SPT | shallower pool (pay) test |
| sptd | spotted |
| sptty | spotty |
| Spud Date | date actually started drilling |

| | |
|---|---|
| SPWY | spillway |
| sq | square |
| sq cg | squirrel cage |
| sq cm | square centimeter |
| sq ft | square foot |
| sq in | square inch |
| sq km | square kilometer |
| sq m | square meter |
| sq mm | square millimeter |
| sq pkr | squeeze packer |
| sq yd | square yard (s) |
| SQRT | square root |
| sqz | squeeze (ed) (ing) |
| SR | short radius |
| SR | swab rate |
| SR | swab run (s) |
| SRB | sulfate bacteria |
| SRG | surge |
| SRL | single random lengths |
| SRN | straight-run naphtha |
| srt | sort (ed) (ing) |
| SRV | safety relief valve |
| SS | sandstone |
| SS | service station |
| SS | shock sub |
| SS | short string |
| SS | single shot |
| SS | slow set (cement) |
| SS | small show |
| SS | stainless steel |
| SS | string shot |
| SS | subsea |
| SS | subsurface |
| SSA | spot sales agreement |

## Abbreviations with Definitions

| | |
|---|---|
| SSG | slight show of gas |
| SSO | slight show of oil |
| SSO&G | slight show of oil and gas |
| SSSV | subsurface safety valve |
| SSU | Saybolt Seconds Universal |
| SSUW | salty sulfur water |
| ST | short thread |
| st | start |
| ST (g) | sidetrack (ing) |
| St L | Saint Louis lime |
| St Ptr | Saint Peter |
| St Gen | Saint Genevieve |
| ST&C | short threads & coupling |
| sta | station |
| Sta Marg | Santa Margarita |
| stab | stabilized (er) |
| STAG | staggered |
| Stal | Stalnacker |
| Stan | Stanley |
| stat | stationary |
| stat | statistical |
| State pot | state potential |
| stb, STB | stock tank barrels |
| stb/d, STBPD | stock tank barrels per day |
| stcky | sticky |
| std | stand (s) (ing) |
| std (s) (g) | standards |
| stdg | standing |
| stdy | steady |
| stdy | study |
| STDZN | standardization |
| Stel | Steele |
| steno | stenographer |
| Stens | Stensvad |

| | |
|---|---|
| STG | stage |
| stging | straightening |
| STH | sidetracked hole |
| STIF | stiffener |
| stip | stippled |
| stir | stirrup |
| stk | stock |
| stk | streak (s) (ed) |
| stk | stuck |
| stl | steel |
| stm | steam |
| STM | steel tape measurement |
| stm cyl oil | steam cylinder oil |
| stm eng oil | steam engine oil |
| STM TR | steam trace (ing) |
| stn (d) (g) | stain (ed) (ing) |
| Stn Crl | Stone Corral |
| stn/by | stand by |
| stncl (d) (g) | stencil (ed) (ing) |
| Stnka | Satanka |
| stnr | strainer |
| stoip | stock tank oil in place |
| stor | storage |
| STP | standard temperature and pressure |
| STPR | strip (per) (ping) |
| stpr (d) | stopper (rd) |
| Str | Strawn |
| strat | stratigraphic |
| strd | straddle |
| strd | strand (ed) |
| strg | storage |
| strg | strong |
| strg (r) | string (er) |
| stri | striated |

## Abbreviations with Definitions

| | |
|---|---|
| strk | streak |
| STROH | strap out of hole |
| strom | stromatoporoid |
| strt | straight |
| strtd | straightened |
| struc | structure, structural |
| STTD | sidetracked total depth |
| STV | stock tank vapor |
| stv | stove oil |
| stwy | stairway |
| Sty Mtn | Stony Mountain |
| styo | styolite, styolitic |
| sub | subsidiary |
| sub | substance |
| sub angl | subangular |
| Sub Clarks | sub-Clarksville |
| sub rnd | subrounded |
| subd | subdivision |
| SUBST | substitute |
| substa | substation |
| suc | sucrose, sucrosic |
| suct | suction |
| sug | sugary |
| sul | sulfur |
| sul wtr | sulfur water |
| SULF | sulfur, sulfuric |
| sulf | sulfated |
| SUM | summarize |
| sum | summary |
| Sum | Summerville |
| Sunb | Sunburst |
| Sund | Sundance |
| Sup | Supai |
| supl | supply (ied) (ier) (ing) |

| | |
|---|---|
| supp | supplement |
| suppt | support |
| suprv | supervisor |
| supsd | superseded |
| supt | superintendent |
| sur | survey |
| Surp | surplus |
| SUS | Saybolt universal seconds |
| susp | suspended |
| SUSP CLG | suspended ceiling |
| SUV | Saybolt universal viscosity |
| SV | solenoid valve |
| svc | service |
| svcu | service unit |
| SVI | smoke volatility index |
| Svry | severy |
| SW | salt wash |
| SW | salt water |
| SW | socket weld |
| SW | southwest |
| SW | spiral weld |
| SW | switch |
| SW&W | show gas and water |
| SW/c | southwest corner |
| SW/4 | southwest quarter |
| Swas | Swastika |
| SWB | seal-welded bonnet |
| swbd | swabbed, swabbing |
| swbd | switchboard |
| SWC | sidewall cores |
| SWCM | saltwater-cut mud |
| SWD | saltwater disposal |
| swd | swaged |
| SWDS | saltwater disposal system |

SWDW saltwater disposal well
Swet sweetening
SWF saltwater fracture
swgr switchgear
SWI saltwater injection
SWION shut well in overnight
SWLD seal weld
SWNP sidewall neutron porosity
SWP steam working pressure
SWR statewide rules
SWRK switchrack
SWS sidewall samples
SWTS salt water to surface
SWU swabbing unit
sx sacks
sxtu sextuple
Syc Sycamore
Syl Sylvan
sym symbol
sym symmetrical
syn synchronous, synchronizing
syn synthetic
syn conv synchronous converter
SYNSCP synchroscope
SYNTH synthesis
sys system
sz size

# T

T tee
T tooth, teeth
T ton (after number-3T)

| | |
|---|---|
| T | township (as T2N) |
| T&B | top and bottom |
| T&C | threaded and coupled |
| T&C | topping and coking |
| T.O. | temperature observation |
| T&BC | top and bottom chokes |
| T&G | tongue and groove (joint) |
| T&R | tubing and rods |
| T&W | tarred and wrapped |
| T/ | top of (a formation) |
| T/Box | terminal box |
| T/BRD | terminal board |
| T/C | tank car |
| T/C | turbine compressor |
| T/pay | top of pay |
| T/S | top salt |
| T/sd | top of sand |
| TA | temporarily abandoned |
| TA | turn around |
| tab | tabular, tabulating |
| TACH | tachometer |
| Tag | Tagliabue |
| Tal | Tallahatta |
| Tamp | Tampico |
| TAN | tangent |
| Tan | Tansill |
| Tann | Tannehill |
| TAPS | Trans-Alaska Pipeline System |
| Tark | Tarkio |
| Tay | Taylor |
| TB | tank battery |
| TB | thin bedded |
| tb | tube |
| TB/BDL | tube bundle |

| | |
|---|---|
| TBA | tertiary butyl alcohol |
| TBA | tires, batteries, and accessories |
| TBE | threaded both ends |
| tbg | tubing |
| tbg chk | tubing choke |
| TBP | true boiling point |
| TC | temperature controller |
| TC | tool closed |
| TC | top choke |
| TC | tubing choke |
| TCC | tag closed cup (flash) |
| TCC | thermofor catalytic cracking |
| TCC | tubing and casing cutter |
| Tcf, TCF | trillion cubic feet |
| Tcf/d TCF/D | trillion cubic feet per day |
| TCP | tricresyl phosphate |
| TCV | temperature control valve |
| TD | time delay |
| TD | total depth |
| TDA | temporary dealer allowance |
| TDI | temperature differential indicator |
| TDR | temperature differential recorder |
| TDT | thermal decay time |
| TE | temporary |
| tech | technical, technician |
| TEFC | totally enclosed, fan cooled |
| tel | telephone, telegraph |
| TEL | tetraethyl lead |
| Tel Cr | Telegraph Creek |
| Temp | temperature |
| temp | temporary (ily) |
| Tens | Tensleep |
| tens str | tensile strength |
| Tent | tentaculites |

| | |
|---|---|
| tent | tentative |
| Ter | tertiary |
| termin | terminate (ed) (ing) (ion) |
| Tex | Texana |
| tex | texture |
| Text. art. | Textularia articulate |
| Text. d. | Textularia dibollensis |
| Text. h. | Textularia hockleyensis |
| Text. w. | Textularia warreni |
| TFB | trip (ped) for bit |
| Tfing | Three Finger |
| Tfks | Three Forks |
| TFNB | trip for new bit |
| tfs | tuffaceous |
| TG | temperature gradient |
| th | thence |
| TH | tight hole |
| Thay | Thaynes |
| THC | top hole choke |
| THD | thermal hydrodealkylation |
| thd | thread, threaded |
| Ther | Thermopolis |
| therm | thermometer |
| therm ckr | thermal crack |
| therst | thermostat |
| THF | tubinghead flange |
| THFP | top hole flow pressure |
| thk | thick, thickness |
| thrling | throttling |
| thrm | thermal |
| thru | through |
| Thur | Thurman |
| TI | temperature indicator |
| ti | tight |

## Abbreviations with Definitions

| | |
|---|---|
| TIC | temperature indicator controller |
| TIH | trip in hole |
| TIM | Timpas |
| Timpo | Timpoweap |
| tk | tank |
| TKF | tank farm |
| tkg | tankage |
| tkr | tanker (s) |
| tl | tool (s) |
| tl jt | tool joint |
| TLE | thread large end |
| TLG | telegraph |
| TLH | top of liner hanger |
| TLP | tension leg platform |
| TML | tetramethyl lead |
| tndr | tender |
| TNS | tight no show |
| TO | temperature observation |
| TO | tool open |
| TOBE | thread on both ends |
| TOC | tag open cup (flash) |
| TOC | top of cement |
| TOCP | top of cement plug |
| Tod | Todilto |
| TOE | threaded one end |
| TOF | top of fish |
| TOH | trip out of hole |
| tol | tolerance |
| TOL | top of liner |
| tolu | toluene |
| Tonk | Tonkawa |
| tons | tons |
| TOP | testing on pump |
| topg | topping |

| | |
|---|---|
| topo | topographic, topography |
| TOPS | turned over to producing section |
| Tor | Toronto |
| Toro | Toroweap |
| TORT | tearing out rotary tools |
| TOS | top of salt |
| tot | total |
| Tow | Towanda |
| TP | tool pusher |
| TP | Travis Peak |
| TP | treating pressure |
| TP | tbg pressure |
| TP&A | theoretical production and allocation |
| TPC | tubing pressure, closed |
| TPF | threaded pipe flange |
| TPF | tubing pressure, flowing |
| tpk | turnpike |
| Tpka | Topeka |
| TPSI | tubing pressure shut in |
| TR | temperature recorder |
| TR | trace |
| TR | tract |
| trans | transfer (ed) (ing) |
| trans | transformer |
| trans | transmission |
| transl | translucent |
| transp | transparent |
| transp | transportation |
| TRC | temperature recorder controller |
| Tremp | Teremplealeau |
| Tren | Trenton |
| TRG | to be conditioned for gas |
| Tri | Triassic |
| trilo | trilobite |

| | |
|---|---|
| Trin | Trinidad |
| trip | triplicate |
| Trip | Tripoli |
| trip | tripolitic |
| trip | tripped (ing) |
| trk | truck |
| trkg | trackage |
| Trn | Trenton |
| TRNDC | transducer |
| TRO | to be conditioned for oil |
| TRQ | torque |
| trt (r)(d)(g) | treat (er) (ed) (ing) |
| trtr | treater |
| TRVL | travel (ed) (ing) |
| TS | tensile strength |
| TS | topo sheet evaluation |
| TSD | temporarily shutdown |
| TSE | thread small end |
| TSE-WLE | thread small end, weld large end |
| TSI | temporarily shut in |
| TSITC | temperature survey indicated top cement at |
| TSS | Tar Springs sand |
| tst (r) (g) | test (er) (ing) |
| tste | taste |
| TSTM | too small to measure |
| TT | tank truck |
| TT | through-tubing |
| TTC | through-tubing caliber |
| TTF | test to follow |
| TTL | total time lost |
| TTP | through-tubing plug |
| TTT | through the tanks |
| TTTT | turned to test tank |
| Tuck | Tucker |

| | |
|---|---|
| tuf | tuffaceous |
| Tul Cr | Tulip Creek |
| tung carb | tungsten carbide |
| TURB | turbo, turbine |
| Tus | Tuscaloosa |
| TV | television |
| TVA | temporary voluntary allowance |
| TVD | true vertical depth |
| TVP | true vapor process |
| TW | tank wagon |
| Tw Cr | Twin Creek |
| TWI | techniques of water-resources investigations |
| twp | township |
| TWR | tower |
| twst | townsite |
| twst off | twisted off |
| TWTM | too wet (weak) to measure |
| TWX | teletype |
| ty | type |
| typ | typical |
| tywr | typewriter |

## U

| | |
|---|---|
| U | unclassified |
| U/ | upper (i.e., U/Simpson) |
| U/C | under construction |
| U/L | upper and lower |
| U/W | used with |
| U/WTR | utility water |
| UC | upper casing |
| UCH | use customer's hose |
| UD | under digging |

| | |
|---|---|
| UFD | utility flow diagram |
| UG | under gauge |
| UG | underground |
| UGL | universal gear lubricant |
| UHF | ultra-high frequency |
| ULJ | perforating, Ultrajet |
| ult | ultimate |
| UM | Umiat Meridian (Alaska) |
| UMB | umbrella (s) |
| un | unit |
| UNBAL | unbalanced |
| unbr | unbranded |
| UNC | unified coarse thread |
| unconf | unconformity |
| uncons | unconsolidated |
| undiff | undifferential |
| unf | unfinished |
| UNF | unified fine thread |
| uni | uniform |
| Univ | university, universal |
| UNLD | unloading |
| UNLDR | unloader |
| UPS | uninterruptible power supply |
| UR | underreaming |
| UR | unsulfonated residue |
| UR | used rod |
| USG | United States gauge |
| UST | ultrasonic test |
| UT | ultrasonic examination |
| UT | upper tubing |
| UT | upthrown |
| UTL | utility |
| UTM | universal transverse mercator |
| UV | ultraviolet |

| | |
|---|---|
| UV | Union Valley |
| Uvig. lir. | Uvigerina lirettensis |

| | |
|---|---|
| v, V | volt |
| V | volume |
| v. | very (as very tight) |
| v.n. | very noticeable |
| v.c. | very common |
| V.P.S. | very poor sample |
| v.r. | very rare |
| v-f-gr | very fine-grained |
| v-HOCM | very heavily oil-cut mud |
| v-sli | very slight |
| V/DWG | vendor drawing |
| V/L | vapor-liquid ratio |
| V/S | velocity survey |
| v% | volume-percent |
| va | Volt-ampere |
| vac | vacant |
| vac | vacation |
| vac | vacuum |
| Vag. reg. | Vaginuline regina |
| Val | Valera |
| Vang | Vanguard |
| vap (r) | vapor (izor) |
| var | variable, various |
| var | volt-ampere reactive |
| vari | variegated |
| VARN | varnish |
| VCP | vitrified clay pipe |
| vel | velocity |

## Abbreviations with Definitions

| | |
|---|---|
| vent | ventilator |
| Ver Cl | Vermillion Cliff |
| Verd | Verdigris |
| vert | vertical |
| ves | vesicular |
| VESS | vessel |
| vfg | very fine-grain (ed) |
| VGC | viscosity-gravity constant |
| VHF | very high frequency |
| VHGCM | very heavily (highly) gas-cut mud |
| VHGCSW | very heavily (highly) gas-cut salt water |
| VHGCW | very heavily (highly) gas-cut water |
| VHO&GCM | very heavily (highly) oil and gas-cut mud |
| VHO&GCSW | very heavily (highly) oil and gas-cut salt water |
| VHO&GCW | very heavily (highly) oil and gas-cut water |
| VHOCM | very heavily (highly) oil-cut mud |
| VHOCSW | very heavily (highly) oil-cut salt water |
| VHOCW | very heavily (highly) oil-cut water |
| Vi | Viola |
| VI | viscosity index |
| VIB | vibrate (or) (ing) |
| Virg | Virgelle |
| vis | viscosity |
| vis | visible |
| vit | vitreous |
| Vks | Vicksburg |
| VLAC | very light amber cut |
| vlv | valve |
| VM&P Naptha | varnish makers and painters naphtha |
| VOC | volatile organic compounds |
| Vogts | Vogtsberger |
| vol | volume |
| vol% | volume-percent |
| vol. eff. | volumetric efficiency |

| | |
|---|---|
| VOLT | voltage |
| VP | vapor pressure |
| VR | vapor recovery |
| vrs | varas |
| vrtb | vertebrate |
| vrtl | vertical |
| VRU | vapor recovery unit |
| vrvd | varved |
| VS | velocity survey |
| vs | versus |
| VSC | volumetric subcommittee |
| VSGCM | very slightly gas-cut mud |
| VSGCSW | very slightly gas-cut salt water |
| VSGCW | very slightly gas-cut water |
| VSM | vertical support member |
| VSO&GCM | very slightly oil- and gas-cut mud |
| VSO&GCSW | very slightly oil- and gas-cut salt water |
| VSOCM | very slightly oil-cut mud |
| VSOCSW | very slightly oil-cut salt water |
| VSOCW | very slightly oil-cut water |
| VSP | very slightly porous |
| VSSG | very slight show of gas |
| VSSO | very slight show of oil |
| vt | vapor temperature |
| vug | vuggy |
| vug | vugular |

| | |
|---|---|
| W | wall (if used with pipe) |
| W | water-supply paper |
| w | watt |
| W | west |

| | |
|---|---|
| W | wide |
| W Cr | Wall Creek |
| w shd | washed |
| W.O.B. | weight on bit |
| W-F | Washita-Fredericksburg |
| w-hr | watt-hour |
| W&R | wash and ream |
| w/ | with |
| W/CLR | water cooler |
| W/L | water load |
| W/O | west offset |
| W/O | without |
| W/SSO | water with slight show of oil |
| W/S TK | welded steel tank |
| W/sulf O | water with sulfur odor |
| W/2 | west half |
| w% | weight-percent |
| Wab | Wabaunsee |
| WAB | weak air blow |
| WACOG | Weighted Average Cost of Gas |
| WACT | weight averaged catalyst temperature |
| Wad | Waddell |
| WAG | water-alternating gas (or water and gas) |
| Wap | Wapanucka |
| War | Warsaw |
| Was | Wasatch |
| WaSd | Waltersburg sand |
| Wash | Washita |
| WB | water blanket |
| WB | wet bulb |
| WB | Woodbine |
| WBIH | went back in hole |
| WC | water closet |
| WC | water cushion (DST) |

| | |
|---|---|
| WC | water cut |
| WC | wildcat |
| WC | Wolfe City |
| WCM | water-cut mud |
| WCO | water-cut oil |
| WCTS | water cushion to surface |
| WD | water depth |
| WD | water disposal well |
| WD | wiring diagram |
| Wd R | Wind River |
| Wdfd | Woodford |
| WE | weld ends |
| Web | Weber |
| Well | Wellington |
| WF | new field wildcat, dry |
| WF | waterflood |
| WF | wide flange |
| WFD | new field wildcat, discovery |
| WFD | wildcat field, discovery |
| WH | wellhead |
| Wh Dol | white dolomite |
| Wh Sd | white sand |
| WHIP | wellhead injection pressure |
| whip | whipstick |
| WHL | wheel |
| whse | warehouse |
| whsle | wholesale |
| wht | white |
| Wl | washing in |
| Wl | water injection |
| Wl | working interest |
| Wl | wrought iron |
| Wich Alb | Wichita Albany |
| Wich. | Wichita |

## Abbreviations with Definitions

| | |
|---|---|
| WIH | water in hole |
| Willb | Willberne |
| Win | Winona |
| Winf | Winfield |
| Wing | Wingate |
| Winn | Winnipeg |
| WIP | work in place |
| WIW | water injection well |
| wk | weak |
| wk | week |
| wkd | worked |
| wkg | working |
| wko | workover |
| wkor | workover rig |
| WL | water loss |
| WL | well lines |
| WL | west line |
| WL | wireline |
| wlbr | wellbore |
| WLC | wireline coring |
| wld | welded, welding |
| WLD/DET | welding detail (s) |
| wldr | welder |
| WLT | wireline test |
| WLTD | wireline total depth |
| W'ly | westerly |
| WN | weldneck |
| WNSO | water not shut off |
| WO | waiting on |
| WO | wash oil |
| WO | wash over |
| wo | washout |
| WO | wildcat outpost, dry |
| WO | work order |

143

| | |
|---|---|
| WO | workover |
| WOA | waiting on acid |
| WOA | waiting on allowable |
| WOB | waiting on battery |
| WOC | waiting on cement |
| WOCR | waiting on completion rig |
| WOCT | waiting on cable tools or completion tools |
| WODP | waiting on drillpipe |
| WOE | successful wildcat outpost |
| WOG | waiting on geologist |
| WOG | water oil or gas |
| Wolfc | Wolfcamp |
| WOO | waiting on orders |
| Wood | Woodside |
| Woodf | Woodford |
| WOP | waiting on permit |
| WOP | waiting on pipe |
| WOP | waiting on plastic |
| WOP | waiting on pump |
| WOPE | waiting on production equipment |
| WOPL | waiting on pipeline |
| WOPT | waiting on potential test |
| WOPU | waiting on pumping unit |
| WOR | waiting on rig or rotary |
| WOR | water-oil ratio |
| WORT | waiting on rotary tools |
| WOS | washover string |
| WOSP | waiting on state potential |
| WOST | waiting on standard tools |
| WOT | waiting on test or tools |
| WOT&C | waiting on tank and connection |
| WOW | waiting on weather |
| WP | new pool wildcat, dry |
| WP | wash pipe |

## Abbreviations with Definitions

| | |
|---|---|
| WP | well pad |
| WP | working pressure |
| WPD | new pool wildcat, discovery |
| WPM | well pad manifolding |
| wpr | wrapper |
| WPT | Windfall Profit Tax |
| WR | White River |
| Wref | Wreford |
| WRG | wiring |
| WRTB | wash and ream to bottom |
| WS | shallower pool wildcat, dry |
| WS | water saturation |
| WS | whipstock |
| WS | worldscale |
| WSD | shallower pool wildcat, discovery |
| WSD | whipstock depth |
| wsh (g) | wash (ing) |
| WSIM | water separation index modified |
| WSO | water shutoff |
| WSONG | water shutoff no good |
| WSOOK | water shutoff OK |
| WST | waste |
| WSW | water source wells |
| WSW | water supply well |
| WT | wall thickness (pipe) |
| wt | weight |
| wt% | weight-percent |
| WTB | wash to bottom |
| wtg | waiting |
| WTH/PRF | weatherproof |
| wthr(d) | weather (ed) |
| wtr (y) | water, watery |
| wtr. cush | water cushion |
| WTR/PRF | waterproof |

Standard Oil & Gas Abbreviator

| | |
|---|---|
| WTR/T | watertight |
| WTS | water to surface |
| WUT | water-up to |
| WW | wash water |
| WW | water well |
| Wx | Wilcox |

| | |
|---|---|
| X | salt |
| X-bdd(ing) | crossbedded, crossbedding |
| X-hvy | extra heavy |
| X-line | extreme line |
| X-over | crossover |
| X-R | X-ray |
| X-REF | cross reference |
| X-SECT | cross section |
| x-stg | extra strong |
| x/n | crystalline |
| XFMR | transformer |
| XHGR | extra-heavy grade pipe |
| Xing | crossing |
| Xlam | cross-laminated |
| Xln | crystalline |
| XMTR | transmitter |
| XO | crossover |
| XO-sub | crossover sub |
| xtal | crystal |
| Xtree | Christmas tree |
| XW | salt water |
| XX-Hvy | double extra heavy |
| XX-STR | double extra strong |

## Y

| | |
|---|---|
| y | Yates |
| yd | yard (s) |
| yel | yellow |
| YIL | yellow indicating lamp |
| YMD | your message of date |
| YMY | your message yesterday |
| Yoak | Yoakum |
| YP | yield point |
| yr | year |
| Yz | Yazoo |

## Z

| | |
|---|---|
| zen | zenith |
| Zil | Zilpha |
| ZN | zinc |
| Zn | zone |

## Miscellaneous

| | |
|---|---|
| 10¹² | trillion |
| 12 GA W.W.S. | 12 gauge wire-wrapped screen (in a liner) |
| 3 PH | three phase |
| 3P ST SW | triple pole single throw switch |
| 3P SW | triple pole switch |
| 4P ST SW | four sole single throw switch |
| 4P SW | four pole switch |
| 8rd | eight round pipe |
| /ft | per foot |
| /L | line, as in E/L (eastline) |
| % | percent |
| °API | degrees, API |
| °C | degrees Centrigrade, degrees Celcius |
| °F | degrees Fahrenheit |
| µg | microgram (s) |

Definitions with Abbreviations

## A

| | |
|---|---|
| abandoned | abd |
| abandon-salvage deferred | ASD |
| abandoned gas well | abd-gw |
| abandoned location | abd loc |
| abandoned oil & gas well | abd-ogw |
| abandoned oil well | abd-ow |
| about | abt |
| above | abv |
| abrasive jet | abrsi jet |
| absolute | abs |
| absolute bottom-hole location | ABHL |
| absolute open flow potential (gas well) | AOF |
| absorber | asbr |
| absorption | absm |
| abstract | abst |
| abstract (i.e., A-10) | A |
| abundant | abun |
| accelerometers | ACCEL |
| access | ACC |
| accessory, accessories | ACCESS |
| account(ing) | acct |
| accounts receivable | A/R |
| accumulative, accumulator | accum |
| acid | ac |
| acid frac | AF |
| acid fracture treatment | acfr |
| acid residue | AR |
| acid-soluble oil | ASO |
| acid treat (ment) | AT |
| acid water | AW |
| acid-cut mud | ACM |

149

| | |
|---|---|
| acid-cut water | ACW |
| acidity or alkalinity | pH |
| acidize (ed) (ing) | acd |
| acidized with | A/ |
| acoustic caliper | CMA |
| acoustic cement | A-Cem |
| acre feet | ac-ft |
| acre(s), acreage | ac |
| acreage | ac, acrg |
| acre-feet | ac-ft |
| acrylonitrile butadiene styrene rubber | ABS |
| actual | ACT |
| actual drilling | AD |
| actual drilling cost | ADC |
| actual drilling time | ADT |
| actual jetting time | AJT |
| actuated, actuator | ACT |
| adapter | adpt |
| addition or modification reque | AMR |
| additional | addl |
| additive | add |
| adhesive | ADH |
| adjustable | adj |
| adjustable spring wedge | ASW |
| adjustments and allowances | A&A |
| administration, administrative | adm |
| adomite | ADOM |
| adsorption | adspn |
| advanced | advan |
| aeration, aerator | AER |
| affidavit | afft |
| affirmed | affd |
| after acidizing, as above | AA |
| after condenser | AF/COND |

| | |
|---|---|
| after cooler | AF/CLR |
| after federal income tax | AFIT |
| after fracture | AF |
| after receipt of order (purchasing term) | ARO |
| after shot | AS |
| after the tanks | ATT |
| after top center | ATC |
| after treatment | AT |
| agglomerate | agim |
| aggregate | AGGR |
| agitator | AG |
| air conditioning | A/C |
| air conditioning | AIR COND |
| air cooled | A/CLD |
| air cooler | A/CLR |
| air quality control region | AQCR |
| air quality maintenance area | AQMA |
| alarm | alm |
| Alaskan North Slope | ANS |
| Albany | Alb |
| alcoholic | alc |
| algae | alg |
| alignment (ing) | ALIGN |
| alkaline, alkalinity | alk |
| alkalinity or acidity | pH |
| alkylate, alkylation | alkyl |
| all thread | AT |
| allocation | ALOC |
| allowable not yet available | ANYA |
| allowable, allowance | ALLOW |
| alloy | ALY |
| along | alg |
| alternate | alt |
| alternating current | AC |

| | |
|---|---|
| altitude | ALT |
| aluminum | AL |
| aluminum conductor steel reinforced | ACSR |
| ambient | amb |
| American Chemical Society | ACS |
| American melting point | AMP |
| American Petroleum Institute | API |
| American Public Health Association | APHA |
| American Society for Testing & Materials | ASTM |
| American Standards Association | ASA |
| American Steel & Wire gauge | AS&W ga |
| American Water Works Association | AWWA |
| American Wire Gauge | AWG |
| ammeter | AMM |
| ammonia | $NH_3$ |
| ammonia chloride | $NH_4Cl$ |
| amorphous | amor |
| amortization | amort |
| amount | amt |
| amount not reported | ANR |
| ampere | amp |
| ampere hour | amp hr |
| amphipore | amph |
| Amphistegina | amph |
| analysis, analytical | anal |
| Anchor (age) | ANC |
| and husband | et con. |
| and husband | et vir. |
| and others | et al. |
| and the following | et seq. |
| and wife | et ux. |
| angle, angular | ang |
| angstrom unit | Å |
| Angulogerina | Angul |

## Definitions with Abbreviations

| | |
|---|---|
| anhydrite stringer | AS |
| anhydrite, anhydritic | anhy |
| anhydrous | anhyd |
| annubar | ANUB |
| annular velocity | AV |
| annulus | ann |
| annunciator | ANUC |
| apartment | apt |
| apparatus | APPAR |
| apparent (ly) | apr |
| appears, appearance | app |
| appliance | appl |
| application | applic |
| applied | appl |
| approved | appd |
| approved total depth | ATD |
| approximate (ly) | approx |
| aqueous | aq |
| aragonite | arag |
| Arapahoe | Ara |
| Arbuckle | Arb |
| Archeozoic | Archeo |
| architectural | arch |
| area of mutual interest | AMI |
| arenaceous | aren |
| argillaceous | arg |
| argillite | arg |
| Arkadelphia | Arka |
| arkose(ic) | ark |
| armature | arm |
| aromatics | arom |
| around | arnd |
| arrange (ed) (ing) (ment) | ARR |
| articles published in outside journals/books | OP |

| | |
|---|---|
| artificial lift | AL |
| as soon as possible | ASAP |
| asbestos | asb |
| ashern | Ash |
| asphalt. asphaltic | asph |
| asphaltic stain | astn |
| assembly | assy |
| assigned | assgd |
| assignment | asgmt |
| assistant | asst |
| associate (d) (s) | assoc |
| association | assn |
| Association of American Railroads | AAR |
| Association of Desk & Derrick Clubs | ADDC |
| Association of Official Agricultural Chemistry | AOAC |
| at rate of | ARO |
| Atlas Bradford modified | ABM |
| atmosphere, atmospheric | atm |
| Atoka | At |
| atomic | at |
| atomic weight | at wt |
| attach (ed) (ing) (ment) | ATT |
| attempt(ed) | att |
| attorney | atty |
| Audit Bureau of Circulation | ABC |
| auditorium | aud |
| Austin | Aus |
| Austin chalk | AC |
| Authorization for Commitment | AFC |
| Authorization for Expenditure | AFE |
| Authorization to Proceed | ATP |
| authorized | auth |
| authorized depth | AD |
| Authorized for Construction | AFC |

## Definitions with Abbreviations

| | |
|---|---|
| automatic | auto |
| automatic custody transfer | ACT |
| automatic data processing | ADP |
| automatic transmission fluid | ATF |
| automatic volume control | AVC |
| automotive | auto |
| automotive gasoline | autogas |
| Aux Vases sand | AV |
| auxiliary | aux |
| auxiliary flow diagram | AFD |
| available | avail |
| average | avg |
| average flowing pressure | AFP |
| average freight rate assessment | AFRA |
| average injection rate | AIR |
| average penetration rate | APR |
| average treating pressure | ATP |
| average tubing pressure | ATP |
| aviation | av |
| aviation gasoline | avgas |
| awaiting | awtg |
| award | AWD |
| azeotrophic | aztrop |
| azimuth | az |

## B

| | |
|---|---|
| back flush | BKFLSH |
| back pressure | BP |
| back-pressure valve | BPV |
| back scuttled | B/S |
| back to back | B/B |
| backed out (off) | BO |
| backwash | BKWSH |

155

| | |
|---|---|
| baffle | BFL |
| bailed | bld |
| bailed dry | B/dry |
| bailer | blr |
| bailer feed water | BFW |
| bailing | blg |
| balance | BAL |
| ball joint | B/JT |
| ball sealers | BS |
| ball valve | B/Vlv |
| Balltown sand | Ball |
| band (ed) | bnd |
| barge deck to mean low water | BD-MLW |
| barite (ic) | bar |
| Barker Creek | Bark Crk |
| Barlow Lime | Bar |
| barometer, barometric | bar |
| barrel | bbl |
| barrel water load | BWL |
| barrels acid | BA |
| barrels acid residue | BAR |
| barrels acid water | BAW |
| barrels acid water per day | BAWPD |
| barrels acid water per hour | BAWPH |
| barrels acid water under load | BAWUL |
| barrels condensate | BC |
| barrels condensate per day | BCPD |
| barrels condensate per hour | BCPH |
| barrels condensate per million | BCPMM |
| barrels diesel oil | BDO |
| barrels distillate | BD |
| barrels distillate per day | BDPD |
| barrels distillate per hour | BDPH |
| barrels fluid | BF |

## Definitions with Abbreviations

| | |
|---|---|
| barrels fluid per day | BFPD |
| barrels fluid per hour | BFPH |
| barrels formation water | BFW |
| barrels frac oil | BFO |
| barrels fresh water | BFW |
| barrels liquid per day | BLPD |
| barrels load | BL |
| barrels load & acid water | BL&AW |
| barrels load condensate | BLC |
| barrels load condensate per day | BLCPD |
| barrels load condensate per hour | BLCPH |
| barrels load oil | BLO |
| barrels load oil per day | BLOPD |
| barrels load oil per hour | BLOPH |
| barrels load oil recovered | BLOR |
| barrels load oil to be recovered | BLOTBR |
| barrels load oil yet to recover | BLOYTR |
| barrels load water | BLW |
| barrels load water per day | BLWPD |
| barrels load water per hour | BLWPH |
| barrels load water to recover | BLWTR |
| barrels mud | BM |
| barrels new oil | BNO |
| barrels new water | BNW |
| barrels oil | BO |
| barrels oil per calendar day | BOPCD |
| barrels oil per day | BOPD |
| barrels oil per hour | BOPH |
| barrels oil per producing day | BOPPD |
| barrels per barrel | B/B |
| barrels per day | BPD |
| barrels per hour | B/hr |
| barrels per minute | B/M |
| barrels per stream day | BPSD |

| | |
|---|---|
| barrels per stream (refinery) | B/SD |
| barrels per well per day | BPWPD |
| barrels pipeline oil | BPLO |
| barrels pipeline oil per day | BPLOPD |
| barrels salt water | BSW |
| barrels salt water per day | BSWPD |
| barrels salt water per hour | BSWPH |
| barrels of water | BW |
| barrels water over load | BWOL |
| barrels water per day | BWPD |
| barrels water per hour | BWPH |
| Bartlesville | Bart |
| basal | bsl |
| Basal Oil Creek Sand | BOCS |
| base | B/ |
| Base Blane | B. Bl |
| base of the salt | B slt |
| base Pennsylvania | BP |
| base plate | BSPL |
| base salt | B/S |
| basement | bsmt |
| basement (granite) | base |
| basic sediment | BS |
| basic sediment and water | BS&W |
| basket | bskt |
| Bateman | Bate |
| battery | btry |
| Baumé | Be |
| beaded and center beaded | B&CB |
| Bear River | Bear Riv |
| bearing | brg |
| Bearpaw | BP |
| Beckwith | Beck |
| becoming | bec |

| | |
|---|---|
| bedding | BDNG |
| before acid treatment | BAT |
| before federal income tax | BFIT |
| before top dead center | BTDC |
| Beldon | Bel |
| belemnites | Belm |
| bell and bell | B&B |
| bell and flange | B&F |
| bell and spigot | B&S |
| Belle City | Bel C |
| Belle Fourche | Bel F |
| benchmark | BM |
| bending schedule | B/S |
| Benoist (Bethel) sand | Ben, BT |
| bent & bowed pipe | B&B |
| Benton | Ben |
| bentonite | Bent |
| benzene | bnz |
| benzene toluenexylene (unit) | BTX |
| Berea | Be |
| between | btw |
| bevel (ed) | bev |
| bevel both ends | BBE |
| bevel large end | BLE |
| bevel one end | BOE |
| bevel small end | BSE |
| beveled for welding | BV/WLD |
| beveled end | B.E. |
| bid summary | BID SUM |
| Big Horn | B. Hn. |
| Big Injun | B. Inj. |
| Big Lime | B. Ls |
| Bigenerina | Big. |
| Bigenerina floridana | Big. f. |

| | |
|---|---|
| Bigenerina humblei | Big. h. |
| Bigenerina nodosaria | Big. nod. |
| bill of lading | B/L |
| bill of material | B/M |
| bill of sale | B/S |
| billion | B |
| billion cubic feet | BCF, Bcf |
| billion cubic feet per day | BCFD, Bcfd |
| billion standard cubic feet | Bscf |
| billion standard cubic feet per day | Bscf/D, Bscfd |
| binary | BIN |
| biochemical oxygen demand | BOD |
| biotite | bio |
| Birmingham (or Stubbs) iron wire gauge | BW ga |
| Birmingham wire gauge | Bwg |
| bitumen | bit |
| bituminous | bit |
| black | blk |
| Black Leaf | Blk Lf |
| Black Lime | Blk Li |
| Black Magic (mud) | BM |
| black malleable iron | BMI |
| Black River | B. Riv |
| black sulfur water | BSUW |
| blank liner | blk lnr |
| blast cabinet | Bl/Cb |
| blast joint | BL/JT |
| bleeding | bld |
| bleeding gas | bld gas |
| bleeding oil | bldo |
| blend (ed) (er) (ing) | BLND |
| blew out | BO |
| blind flange | BLD FLG, BF |
| Blinebry | Blin |

| | |
|---|---|
| block | blk |
| block valve | BV |
| blocked off | BO |
| blossom | Blos |
| blow | blo |
| blowdown | bd |
| blow-down test | BDT |
| blower | BLWR |
| blowout equipment | BOE |
| blowout preventer | BOP |
| blowout preventer equipment | BOPE |
| blue | bl |
| board | bd |
| board foot (feet) | bd ft |
| Bodcaw | Bod |
| body wall loss | BWL |
| boiled water | BW |
| boiler | BLR |
| boiler feed water | BFW |
| boiling point | BP |
| Bois d'Arc | Bd'A |
| Bolivarensis | Bol. |
| Bolivina a. | Bol. a. |
| Bolivina floridana | Bol. flor. |
| Bolivina perca | Bol. p. |
| Bone Spring | BS |
| Bonneterre | Bonne |
| booster | BSTR |
| borehole compensated sonic | BHCS |
| bottom (basic) sediment & water | BS&W |
| bottom (ed) | btm (d) |
| bottom choke | btm chk |
| bottom hole | BH |
| bottom-hole assembly | BHA |

| | |
|---|---|
| bottom-hole choke | BHC |
| bottom-hole flowing pressure | BHFP |
| bottom-hole location | BHL |
| bottom-hole money | BHM |
| bottom-hole orientation | BHO |
| bottom-hole pressure | BHP |
| bottom of given formation (i.e., B/Frio) | B/ |
| bottom sediment | BS |
| bottom settlings | BS |
| bottom-hole pressure bomb | BHPB |
| bottom-hole pressure closet (*see also* SIBHP and BHSIP) | BHPC |
| bottom-hole pressure flowing | BHPF |
| bottom-hole pressure survey | BHPS |
| bottom-hole shutin pressure | BHSIP |
| bottom-hole temperature | BHT |
| boulders | bldrs |
| boundary | bndry |
| box (es) | bx |
| box end | be |
| brace (ing) (ed) | BRC |
| brachiopod | brach |
| brackets(s) | brkt(s) |
| brackish (water) | brksh |
| Bradenhead Flange | BHF |
| brake horsepower | bhp |
| brake horsepower-hour | bhp-hr |
| brake mean effective pressure | BMEP |
| brake specific fuel consumption | BSFC |
| brakes | BRKS |
| break (broke) | brk |
| breakdown | BD |
| breakdown acid | BDA |
| breakdown pressure | BDP |

## Definitions with Abbreviations

| | |
|---|---|
| breaker | BRKR |
| breccia | brec |
| bridge plug | BP |
| bridged back | BB |
| bridger | Brid |
| Brinell hardness number | BHN |
| British Standards Institution | BSI |
| British thermal unit | BTU |
| brittle | brtl, brit |
| broke (break) down formation | BDF |
| broken | brkn |
| broken sand | brkn sd |
| bromide | brom |
| brown | bn |
| brown | brn or br |
| Brown & Sharpe (gauge) | B&S ga |
| brown lime | Brn Li |
| brown oil stain | BOS |
| brown shale | brn sh |
| brownish | brnsh |
| bryozoa | bry |
| Buckner | Buck |
| buck-off | b/off |
| buck-on | b/on |
| buckrange | Buckr |
| budgeted depth | BD |
| buff | bf |
| building | bldg |
| building derrick | bldg drk |
| building rig | BR |
| building roads | bldg rds |
| Buliminella textularia | Bul. text. |
| bulk plant | BP |
| bulk vessel | B/VESS |

| | |
|---|---|
| bull plug | BP |
| bullets | blts |
| Bullwaggon | Bull W |
| bumper | bmpr |
| bundle | BDL |
| Burgess | Burg |
| burner | bunr |
| bushel | bu |
| bushing | BSHG |
| butane and propane mix | BP mix |
| butane-butene fraction | BB fraction |
| butt weld | BW, BTWLD |
| butterfly valve | BRFL/V, BTFL/V |
| buttress thread | butt |
| buttress thread coupling | BTC |
| buzzer | BUZ |
| bypass | BYP |
| bypass cooler | BP/CLR |

## C

| | |
|---|---|
| cabinet | CAB |
| cable (ing) | CBL |
| cable tool measurement | CTM |
| cable tools | CT |
| Caddell | Cadd |
| cadmium plate | CD PL |
| cake | ck |
| calcareous, calcerenite | calc |
| calceneous | cale |
| calcite, calcitic | cal |
| calcium | calc |
| calcium-base grease | calc gr |
| calcium chloride | $CaCl_2$ |

| | |
|---|---|
| calcium oxcide | CaO |
| calculate (ed), calculation | Calc |
| calculated absolute open flow | CAOF |
| calculated open flow (potential) | COF |
| calendar day | CD |
| calibrate (tion) | CALIBR |
| caliche | cal |
| California Coordinate System | CCS |
| caliper log | CAL |
| caliper survey | cal |
| caulking | CLKG |
| calorie | cal |
| Calvin | Calv |
| Cambrian | Camb |
| Camp Colorado | Cp Colo |
| Cane River | Cane Riv |
| canvas-lined metal petal basket | CLMP |
| canyon | cany, cyn |
| capacity, capacitor | cap |
| Captain | Cap |
| carbon copy | CC |
| carbon dioxide | $CO_2$ |
| carbon disulfide | $CS_2$ |
| carbon monoxide | CO |
| carbon oxygen | CO |
| carbon residue (Conradson) | CR Con |
| carbon steel | CS |
| carbontetrachloride | carb test |
| carbonaceous | carb |
| carburetor air temperature | CAT |
| care of | c/o |
| Carlile | Car |
| carload | CL |
| Carmel | Carm |

| | |
|---|---|
| Carrizo | Cz |
| carton | ctn |
| cased hole | C/H |
| cased reservoir analysis | CRA |
| casing | csg |
| casing cemented (depth) | CC |
| casing choke | Cck |
| casing collar locator | CCL |
| casing collar perforating record | CCPR |
| casing flange | CF |
| casing point | csg pt, CP |
| casing pressure | csg press, CP |
| casing pressure, shut in | CPSI |
| casing pressure, closed | CPC |
| casing pressure, flowing | CPF |
| casing seat | CS |
| casing set at | CSA |
| casinghead flange | CHF |
| casinghead gas | CHG |
| casinghead pressure | CHP |
| Casper | Casp |
| cast carbon steel | CCS |
| cast iron | CI |
| cast steel | CS |
| cast-iron bridge plug | CIBP |
| cat-cracked light gas oil | CCLGO |
| Cat Creek | Cat Crk |
| Catahoula | Cat |
| catalog | CAT |
| catalyst, catalytic | CAT |
| catalytic cracker | Cat ckr |
| catalytic cracking unit | CCU |
| cathode ray tube | CRT |
| cathodic | cath |

| | |
|---|---|
| Cattleman | Ctlmn |
| caustic | caus |
| caving (s) | cvg(s) |
| cavity | cav |
| Cedar Mountain | Cdr Mtn |
| cellar | cell |
| cellar & pits | C&P |
| cellular | cell |
| Celsius | C |
| cement (ed) | cem |
| cement (ed) (ing) | cmt (d) (g) |
| cement dump bailer | CDB |
| cement evaluation | CET |
| cement friction reducer | CFR |
| cement friction retarder | CFR |
| cement in place | CIP |
| cement to surface | CTS |
| cemented through perforations | cp's |
| Cenozoic | Ceno |
| center (d) | cntr |
| center (land description) | C |
| center line | C/L |
| center of casinghead flange | CCHF |
| center of gravity | CG |
| center of tubing flange | CTH |
| center section line | CSL |
| center to center | C to C |
| center to end | C to E |
| center to face | C to F |
| centigrade | C |
| centigram | cg |
| centiliter | cl |
| centimeter | cm |
| centimeter-gram-second-system | cgs |

| | |
|---|---|
| centimeters per second | cm/sec |
| centipose (s) | cp |
| centistokes | cs |
| central compressor plant | CCP |
| central delivery point | CDP |
| central processing facility | CPF |
| centralizers | cent |
| centrifugal | centr |
| centrifuge | cntf |
| cephalopod | ceph |
| Ceratobulimina eximia | Cert. ex. |
| certified | CERT |
| certified drawing outline | CDO |
| certified public accountant | CPA |
| cetane number | CN |
| chain operated | CH OP |
| chairman | chrm |
| chalcedony | chal |
| chalk | chk |
| chalky | chky |
| chamber | CHMBR |
| chamfer | CHAM |
| change (ed) (ing) | chng |
| changed (ing) bits | CB |
| changed drillpipe | chngd DP |
| channel | CHNL |
| Chappel | Chapp |
| characteristics | CHAR |
| charge (ed) (ing) | chrg (d) (ing) |
| Charles | Char |
| chart | cht |
| Chattanooga shale | Chatt |
| check | ck |
| check valve | CHKV |

## Definitions with Abbreviations

| | |
|---|---|
| checked | chkd |
| checkerboard | Chkbd |
| checkered plate | CHKD PL |
| chemical oxygen demand | COD |
| chemical products | chem. prod |
| chemical, chemist, chemistry | chem. |
| chemically pure | cp |
| chemically retarded acid | CRA |
| Cherokee | Cher |
| chert | cht |
| cherty | chty |
| Chester | Ches |
| chicksan | cksn |
| Chimney Hill | Chim H |
| Chimney Rock | Chim R |
| Chinle | Chin |
| chitin (ous) | chit |
| chloride (s) | chl |
| chlorinator | CHLR |
| chlorine | $Cl_2$ |
| chlorine long | chl log |
| chloritic | chl |
| choke | ch |
| choke | chk |
| Chouteau lime | Chou |
| Christmas Tree | Xtree |
| Christellaria | Cris |
| chromatograph | chromat |
| chrome molybdenum | cr moly |
| chromium | chrome |
| Chugwater | Chug |
| Cibicides | Cib. |
| Cibicides hazzardi | Cib. h. |
| Cimarron | Cima |

| | |
|---|---|
| circle | cir |
| circuit | cir |
| circular | cir |
| circular mills | cir mils |
| circulate & condition | C&C |
| circulate and reciprocate | C&R |
| circulate bottoms up | CBU, ccBU |
| circulate (ing) (tion) | circ |
| circulated and conditioned mud | C&CM |
| circulated and conditioned hole | C&CH |
| circulated out | CO |
| circumference | CRCMF |
| Cisco | Cis |
| Clagget | Clag |
| Claiborne | Claib |
| clarifier | CLFR |
| Clarksville | Clarks |
| class | CL |
| classification | CLASS |
| clastic | clas |
| Clavalinoides | Clav |
| clay filled | CF |
| Claystone | clyst |
| Clayton | Clay |
| Claytonville | Clay |
| clean out | CO |
| clean out & shoot | CO&S |
| clean up | CU |
| clean (ed) (ing) | cln (d) (g) |
| cleaned out to total depth | COTD |
| cleaning out, cleaned out | CO |
| cleaning to pits | CTP |
| clear, clearance | clr |
| Clearfork | Clfk |

## Definitions with Abbreviations

| | |
|---|---|
| clearing | clrg |
| Cleveland | Cleve |
| Cleveland open cup | COC |
| Cliff House | Cliff H |
| clockwise | cw |
| closed | clsd |
| closed cup | CC |
| closed hole | CH |
| closed hydrocarbon drain | CHD |
| closed-in pressure | CIP |
| Clovery | Clov |
| Coal Bed Methane | CBM |
| coarse crystalline | crs-xln |
| coarse grained | C-gr |
| coarse (ly) | crs, c |
| coat and wrap (pipe) | C&W |
| coated | ctd |
| Cockfield | Cf |
| Coconino | Coco |
| Codell | Cod |
| Cody (Wyoming) | Cdy |
| coefficient | coef |
| coiled tubing unit | CTU |
| coke oven gas | COG |
| cold drawn | CD |
| cold finished | CF |
| cold rolled | CR |
| cold-rolled steel | CRS |
| cold water equivalent | CWE |
| cold working pressure | CWP |
| Coleman Junction | Cole Jct |
| collar | COL |
| collect (ed) (ing) (ion) | coll |
| collector | CLTR |

| | |
|---|---|
| Color American Standard Test Method | Col ASTM |
| colored | COL |
| column | COL |
| Comanche | Com |
| Comanche Peak | Com Pk |
| comanchean | Cmchn |
| Comatula | Com |
| combined, combination | comb |
| combustion | COMB |
| coming out of hole | COOH |
| commenced | comm |
| comment | COMT |
| commercial | coml |
| commission | comm |
| commission agent | C/A |
| commissioner | commr |
| common | com |
| Common Business-Oriented Language | COBOL |
| common data base | CDB |
| common data base task force | CDBTF |
| communication | comm |
| community | comm |
| commutator | COMUT |
| compact | cmpt |
| companion flange bolt and gasket | CFB&G |
| companion flange one end | CFOE |
| companion flanges bolted on | CFBO |
| company | Co |
| company operated | Co. Op. |
| company-operated service stations | Co. Op. S.S. |
| comparator | CMPARTR |
| compartment | compt |
| compensated neutron log | CNL |
| complete (ed) (tion) | comp |

| | |
|---|---|
| completed natural | comp nat |
| completed on pump | C.O.P. |
| complete with | C/W |
| components | COMPT |
| components | compnts |
| compound | CMPD |
| compression and absorption plant | C&A |
| compression-ignition engine | CI engine |
| compression ratio | CR |
| compressor | compr |
| compressor station | compr st |
| computer | COMPTR |
| computer control system | CCS |
| concentrate | conc |
| concentric | cncn |
| concentric | conc |
| conchoidal | conch |
| conclusion | concl |
| concrete | conc |
| condensate | cond |
| condensate-cut mud | CCM |
| condenser | cdsr |
| condition (ed) (ing) | cond |
| conductivity | condt |
| conductor (pipe) | condr |
| conduit | CND |
| confidential | conf |
| confirm (ed) (ing) | conf |
| confirming telephone order (purchasing term) | CTO |
| confirming telephone order (purchasing term) | CPO |
| conflict | confl |
| conglomerate, conglomeritic | cglt |
| conglomerate (itic) | cong |
| connection | conn |

| | |
|---|---|
| conodonts | cono |
| Conradson carbon residue | CCR |
| conserve, conservation | consv |
| consolidated | con |
| consolidated | consol |
| constant | const |
| construction | const |
| consumer tank car | CTC |
| consumer tank wagon | CTW |
| contact caliper | C-Cal |
| container | cntr |
| containment | CNTN |
| contaminated, contamination | contam |
| continue (ed) | cont (d) |
| continuous blowdown | CB |
| continuous directional service | CDS |
| continuous flowmeter | CFM |
| continuous weld | CW |
| contour interval (map) | CI |
| contract depth | CD |
| contractor | contr |
| contractor (i.e., C/John Doe) | C/ |
| contractor furnished equipment | CFE |
| contractor responsibility | contr resp |
| contribution | contrib |
| control (s) | cntl |
| control building | CN/BD |
| control valve | CV |
| controller | cntr |
| convector, convection | CONVT |
| converse | conv |
| convert (er) (ed) | CVTR |
| conveyor | cnvr |
| cooler | CLR |

## Definitions with Abbreviations

| | |
|---|---|
| cooling water | CW |
| cooling water return | CWR |
| cooling water supplying | CWS |
| cooling tower | CT |
| cooling tower | CLG/TWR |
| cooperative | co-op |
| coordinate | coord |
| Coordinating Research Council Inc. | CRC |
| Cooper River Meridian (Alaska) | CPR |
| coquina | coq |
| core barrel | CB |
| core hole | C |
| core (ed) (ing) | cr (d) (g) |
| cored | crd |
| coring | cg |
| corner | cor |
| corner | CNR |
| corporation | Corp |
| correct (ed) (ion) | corr |
| corrected gravity | CG |
| corrected total depth | CTD |
| correlation | correl |
| correspondence | corres |
| corrosion | corr |
| corrosion allowance | CA |
| corrugated | corr |
| cost and freight | C&F |
| cost insurance and freight | C.I.F. |
| cost per gallon | CPG |
| Cottage Grove | Cott G |
| Cotton Valley | CV |
| Council Grove | Counc G |
| counter electromotive force | CEMF |
| counter weight | CNT WT |

| | |
|---|---|
| counterbalance (pumping equip.) | CB |
| counterclockwise | ccw |
| county | Cnty |
| county school lands | CSL |
| coupling | cplg |
| cover | CVR |
| Cow Run | CR |
| cracker | Crkr |
| cracking | crkg |
| cradle (s) | CRDL |
| crane | CRN |
| crawl beam | C/BM |
| creek | crk |
| crenulated | cren |
| Cretaceous | Cret |
| crinkled | crnk |
| crinoid (al) | Crin |
| Cristellaria | Cris |
| critical | crit |
| critical compression pressure | CCP |
| critical compression ratio | CCR |
| Cromwell | Crom |
| cross | CRS |
| crossbedded | crbd |
| crossover | CX |
| crown block | crn blk |
| crude oil | CO |
| crude oil purchasing | COP |
| cryptocrystalline | crypto-xln |
| crystalline | cryst |
| cubic | cu |
| cubic centimeter | cu cm |
| cubic centimeter | cc |
| cubic feet gas | cfg |

## Definitions with Abbreviations

| | |
|---|---|
| cubic feet gas per day | cfgd |
| cubic feet gas per hour | cfgh |
| cubic feet per barrel | cu ft/bbl |
| cubic feet per day | cfd |
| cubic feet per minute | cu ft/min, CFM |
| cubic feet per pound | CFP |
| cubic feet per second | cu ft/sec, CFS |
| cubic foot | cf |
| cubic inch | cu in. |
| cubic meter | cu m |
| cubic meters per day | $m^3/d$ |
| cubic yard | cu yd |
| cubical | CUB |
| culvert | culv |
| cumulative | cum |
| Curtis | Cur |
| curve | CRV |
| cushion | cush |
| customer | CUST |
| cut across grain | CAG |
| cut bank | Cut B |
| cut drilling line | CDL |
| cutback | cutbk |
| Cutler | Cutl |
| cutting oil | Cut Oil |
| cutting oil soluble | Cut Oil Sol |
| cutting oil active-sulfurized-dark | Cut Oil Act Sul-dk |
| cutting oil active-sulfurized-transparent | Cut Oil Act Sul-transpt |
| cutting oil straight mineral | Cut Oil St Mrl |
| cuttings | ctg(s) |
| cyclamina | Cyc |
| Cyclamina cancellata | Cycl canc. |
| cycles per minute | cpm |
| cycles per second | cps |

177

| | |
|---|---|
| cyclone | CYC |
| cyclone | cycl |
| cylinder | cyl |
| Cypress Sand | Cy Sd |
| cypridopsis | cyp. |

# D

| | |
|---|---|
| daily allowable | DA |
| daily average injection barrels | DAIB |
| Dakota | Dak |
| damper | dmpr |
| Dantzler | Dan |
| dark | dk |
| dark brown oil | DBO |
| dark brown oil stains | DBOS |
| Darwin | Dar |
| data processing | DP |
| data sheet | D/S |
| date actually started drilling | spud date |
| date of first production | DFP |
| datum | dat, DM |
| datum faulted out | DFO |
| davit | DVT |
| day | D |
| day to day | D/D |
| days since spudded | DSS |
| dead | dd |
| dead oil show | DOS |
| deadweight tester | DWT |
| deadweight tons | DWT |
| Deadwood | Deadw |
| deaerator | deaer |
| dealer | dlr |

| | |
|---|---|
| dealer tank wagon | DTW |
| deasphalting | deasph |
| debutanizer | debutzr |
| Deca | D |
| Decatherm | DTH |
| decibel | db |
| decigram | dg |
| deciliter | dl |
| decimal | dec |
| decimeter | dm |
| decline | decl |
| decrease (ed) (ing) | decr |
| Deemed Heating Value | DHV |
| deep pool test | DPT |
| deepen | dpn |
| deepening | dpg |
| deethanizer | deethzr |
| deflection | defl |
| Degonia | Deg |
| degree day | DD |
| degree (s) | deg |
| degrees API | oAPI |
| degrees Centigrade | oC |
| degrees Fahrenheit | oF |
| deisobutanizer | deisobut |
| Del Rio | Del R |
| Delaware | Dela |
| Delaware River Area Petroleum Refineries | DRAPR |
| delayed coker | DC |
| delivery (ed) (ability) | delv |
| delivery point | delv pt |
| demand meter | DM |
| demolition | dml |
| demurrage | demur |

| | |
|---|---|
| dendrite (ic) | dend |
| dense | ds, dns |
| density log | D/L, DENL |
| department | dept |
| Department of Energy | DOE |
| depletion | depl |
| depreciation | deprec |
| depropanizer | deprop |
| depth | dpt |
| depth bracket allowable | DBA |
| depth recorder | dpt rec |
| derrick | drk |
| derrick floor | DF |
| derrick floor elevation | DFE |
| Des Moines | Des M |
| deslater | desalt |
| description | desc |
| Desert Creek | Des Crk |
| design | dsgn |
| Desk and Derrick | D&D |
| desorbent | desorb |
| destination | dstn |
| desulfurizer | desulf |
| desuperheater | DSUPHTR |
| detail (s) | det |
| detector | det, DCTR |
| detergent | deterg |
| detrital | detr, dtr |
| develop | DVL |
| develop (ed) (ment) | devel |
| development | D |
| development gas well | DG |
| development oil | DO |
| development oil well | DO |

## Definitions with Abbreviations

| | |
|---|---|
| development redrill (sidetrack) | DR |
| development well, carbon dioxide | DC |
| development well, helium | DH |
| development well, sulfur | DSU |
| development well workover | DX |
| deviate, deviation | dev |
| deviation degrees | DD |
| Devonian | Dev |
| dew point | DP |
| dewatering | DWTR |
| dewaxing | dewax |
| Dexter | Dext |
| diagonal | diag |
| diagram | diag |
| diameter | dia |
| diamond bit | DB |
| diamond core | DC |
| diamond core bit | DCB |
| diaphragm | diaph |
| dichloride | dichlor |
| dichloro diphenyltrichloroethane | DDT |
| diesel (oil) | dsl |
| diesel fuel | DF |
| diesel hydrogen desulfurization | DHDS |
| diesel index | D.I. |
| Diesel No. 2 | D-2 |
| diesel oil well | DOC |
| diethanolamine | DEA |
| diethanolamine unit | DEA unit |
| diethylene | diethyl |
| different (ial) (ence) | diff |
| differential pressure | D/P |
| differential valve (cementing) | DV |
| digging cellar | DC |

| | |
|---|---|
| digging cellar and slush pits | DCLSP |
| digging slush pits | DSP |
| digital | DGTL |
| diglycolamine | DGA |
| diluted | dilut |
| dimension | dim |
| dimethyl sulfide | DMS |
| diminish (ing) | dim |
| Dinwoody | Din |
| dipmeter | DM |
| direct (tion) (tor) | dir |
| direct current | DC |
| directional drilling | dir drlg |
| directional survey | dir sur, DS |
| dirty water disposal | DWD |
| discharge | disch |
| Discorbis | Disc. |
| Discorbis gravelli | Disc. grav. |
| Discorbis normada | Disc. norm. |
| Discorbis yeguaensis | Disc. y. |
| discount | disc |
| discover (ed) (ing) (ion) | disc |
| discovery allowable requested | DAR |
| dismantle | disman |
| dismantle (ing) | dsmt (g) |
| displaced, displacement | displ |
| disseminated | dism |
| distance | dist |
| distillate | dstl |
| distillate-cut mud | DCM |
| distillate, distillation | dist |
| distribute (ed) (ing) (ion) | distr |
| distributed control system | DCS |
| district | dist |

| | |
|---|---|
| ditto | do |
| division | div |
| division office | D/O |
| division order | D.O. |
| dock operating building | DOBLDG |
| Dockum | Doc |
| doctor-treatment | doc-tr |
| document | doc |
| documentation | DOCREQ |
| dogleg severity | DLS |
| doing business as | d/b/a |
| dolomite (ic) | dolo |
| dolstone | dolst |
| domestic | dom |
| domestic airline | dom AL |
| domestic water | DOM WTR |
| Dornick Hills | Dorn H |
| Dothan | Doth |
| double | DBL |
| double end | DE |
| double extra heavy | XX-Hvy |
| double extra strong | XX-STR |
| double hub | DH |
| double pipe | DP |
| double pole double base (switch) | DPDB |
| double pole double throw switch | DPDT SW |
| double pole single base (switch) | DPSB |
| double pole single throw switch | DPST SW |
| double pole switch | DP SW |
| double random lengths | DRL |
| Douglas | Doug |
| down | dn |
| downthrown | DT |
| dozen | doz |

| | |
|---|---|
| draft gauge | DG |
| drain | dr |
| drainage | drng |
| drawing | DWG |
| drawworks | dwks |
| dressed dimension four side | d-d-4-s |
| dressed dimension one side and one edge | d-d-1-s-1-e |
| dressed four sides | d-4-s |
| dressed one side | d-1-s |
| dressed two sides | d-2-s |
| Dresser Atlas | DA |
| drier, drying | dry |
| drift angle | DA |
| drill | drl |
| drill (ed) (ing) out | DO |
| drill (ed) (ing) plug | D/P |
| drill and complete | D&C |
| drill collar | DC |
| drill floor | DF |
| drillpipe | DP |
| drillpipe measurement | DPM |
| drillpipe unloaded | DPU |
| drillsite | DS |
| drillsite facility | DSF |
| drillstem | DS |
| drillstem test | DST |
| drilled | drld |
| drilled-out cement | DOC |
| drilled-out depth | DOD |
| drilled-out plug | DOP |
| driller | drlr |
| driller's top | D/T |
| driller's total depth | DTD |
| drilling | drlg |

| | |
|---|---|
| drilling and well completion | DWC |
| drilling break | DB |
| drilling deeper | DD |
| drilling line | DL |
| drilling mud | DM |
| drilling suspended indefinitely | DSI |
| drilling tender | D/T |
| drilling time | DT |
| drilling with air | DWA |
| drilling with gas | DWG |
| drilling with mud | DWM |
| drilling with oil | DWO |
| drilling with salt water | DWSW |
| drillstem test with straddle packers | DST (Strd) |
| drive (ing) (er) | DRV (R), dr |
| dropped | dropd |
| drum | dr, DRM |
| druse | dr |
| drusy | drsy |
| dry and abandoned | D&A |
| Dry Creek | Dr Crk |
| dry desiccant dehydrator | DDD |
| dry film thickness | DFT |
| dry gas | DG |
| dry-hole contribution | DHC |
| dry hole cost | DHC |
| dry hole drilled deeper | DHDD |
| dry-hole money | DHM |
| dry hole reentered | DHR |
| dual | D |
| dual (double) wall packer | DWP |
| dual injection focus log | DIFL |
| dual lower tubing | DLC |
| dual lower tubing | DLT |

| | |
|---|---|
| dual upper casing | DUC |
| dual upper tubing | DUT |
| dually completed | DC |
| Duck Creek | Dk Crk |
| dumped | DMPD |
| Dun & Bradstreet | D&B |
| Duperow | Dup |
| duplex | dx |
| duplicate | dup |
| duration | DUR |
| Dutcher | Dutch |
| dyna-drilling | DD |
| dynamic | dyn |

# E

| | |
|---|---|
| each | ea |
| Eagle | Egl |
| Eagle Ford | EF |
| Eagle Mills | EM |
| Earlsboro | Earls |
| early well tie-ins | EWT |
| east | E |
| east boundary line | E/BL |
| East Cimarron Meridian (Oklahoma) | ECM |
| east half | E/2 |
| east line | E/L |
| east of Rockies | EOR |
| east of west line | E of W/L |
| east offset | E/O |
| east quarter | E/4 |
| easterly | E'ly |
| Eau Claire | Eau Clr |

## Definitions with Abbreviations

| | |
|---|---|
| eccentric | ECC |
| echinoid | ech |
| economics, economy, economizer | econ |
| Ector (County, TX) | Ect |
| education (tor) | Educ (r) |
| Edwards | Edw |
| Edwards lime | Ed lm |
| effective | eff |
| effective depth | E.D. |
| effective horsepower | EHP |
| efficiency | eff |
| effluent | effl |
| eight round pipe | 8rd |
| ejector | ejtr |
| Elbert | Elb |
| elbow | ell (s) ELB |
| electric accounting machines | EAM |
| electronic log tops | EL/T |
| electric resistance weld | ERW |
| electric weld | EW |
| electric (al) | elec |
| electromagnetic | EMN, Elec/MAG |
| electromagnetic induction | EMNI |
| electromotive force | EMF |
| electron-volts | ev |
| electronic data processing | EDP |
| element, elementary | elem |
| elevation (height) | EL |
| elevation ground | el gr |
| elevation, elevator | elev |
| Elgin | Elg |
| eliminate (tor) (ed) | ELIM |
| Ellenburger | Ellen |
| Ellis-Madison contact | EMS |

| | |
|---|---|
| Elmont | Elm |
| Embar | Emb |
| emergency | emer |
| emergency order | EO |
| emergency shutdown | ESD |
| employee | empl |
| empty container | MT |
| emulsion | emul |
| enamel | enml |
| enclosure | encl |
| end of file | EOF |
| end of line | EOL |
| end of month | EOM |
| end of quarter | EOQ |
| end of year | EOY |
| end point | EP |
| end to end | E/E |
| Endicott | End |
| endothyra | endo |
| engine | eng |
| engineer (ing) | engr (g) |
| Englewood | Eglwd |
| enhanced oil recovery | EOR |
| enlarged | enl |
| Entrada | Ent |
| entrance | ENT |
| entry | ent |
| envelope | ENV |
| environment | ENVIR |
| environmental assessment | EA |
| environmental impact report | EIR |
| environmental impact statement | EIS |
| Eocene | Eoc |
| Eponides | Ep. |

| | |
|---|---|
| Eponides yeguaensis | Ep. y. |
| equal, equalizer | eq |
| equation (before a number) | Eq. |
| equilibrium flash vaporization | EFV |
| equipment | equip |
| equivalent | equiv |
| erection | erect |
| erection mark | ERC/MK |
| Ericson | Eric |
| estate | est |
| estimate (ed) (ing) | est |
| estimated time of arrival | ETA |
| estimated total depth | E.T.D |
| estimated ultimate recovery | EUR |
| estimated yearly consumption | EYC |
| ethane | eth |
| ethylene | ethyle |
| ethylene dichloride | EDC |
| euhedral | euhed |
| European melting point | EMP |
| Eutaw | EU |
| evaluate | eval |
| evaporation, evaporate | evap |
| even sorted | ev-sort |
| examination | EXAM |
| example | EX |
| excavation | exc |
| excellent | excl |
| except | ex |
| exchanger | exch |
| excitation | EXC |
| executive | EXEC |
| executor | Exr |
| executrix | Exrx |

**Standard Oil & Gas Abbreviator**

| | |
|---|---|
| Exeter | Ex |
| exhaust | exh |
| exhibit | exh |
| existing | exist |
| expansion | exp |
| expansion joint | EXP JT |
| expected date of delivery | EDD |
| expendable plug | exp plg |
| expense | exp |
| expire (ed) (ing) (ation) | expir |
| exploratory | E |
| exploratory well | EW |
| exploratory, exploration | expl |
| explosion proof | EX-PRF |
| explosive | explos |
| extended, extension | ext (n) |
| extension manhole | Ext M/H |
| exterior | extr |
| external | ext |
| external upset end | EUE |
| extra heavy | X-hvy |
| extra heavy grade pipe | XHGR |
| extra strong | x-stg |
| extraction | extrac |
| extreme line (casing) | X-line |
| extreme pressure | EP |

## F

| | |
|---|---|
| fabricate (ed) (tion) | fab |
| face of stud | FOS |
| face to face | F to F |
| faced and drilled | F&D |
| facet (ed) | fac |

## Definitions with Abbreviations

| | |
|---|---|
| facility (ies) | FACIL |
| facility capacity limits | FCL |
| failed | fld |
| failure | fail |
| faint | fnt |
| faint air blow | FAB |
| fair | fr |
| Fall River | Fall Riv |
| Farmington | Farm |
| farmout | FO |
| farmout option | F/O opt |
| fasten (ing) (er) | FSTN |
| fault | fkt |
| faulted out | FO |
| fauna | fau |
| favosites | fvst |
| federal | fed |
| Federal Employers Liability Act | FELA |
| Federation of Sewage and Industrial Wastes Association | FSIWA |
| feed | FD |
| feed effluent | FD EFF |
| feed rate | FR |
| feed water | FD/WTR |
| feeder | fdr |
| feedstock | FS |
| feet, foot | ft |
| feet per hour | ft/hr |
| feet per minute | fmp |
| feet per minute | ft/min |
| feet per second | fps |
| feet per second | ft/sec |
| feldspar (thic) | fld |
| female | FEM |

| | |
|---|---|
| female pipe thread | FPT |
| female to female angle | FFA |
| female to female globe (valve) | FFG |
| Ferguson | Ferg |
| ferric sulfate | $Fe_2(SO_4)_3$ |
| ferruginous | ferr |
| Ferry Lake anhydrite | FLA |
| fertilizer | fert |
| fiberglass | fbrgs |
| fiberglass-reinforced plastic | FRP |
| fibrous | fib |
| field | fld |
| field authorized to commence operations | FACO |
| field fabricated | F/FAB |
| field fuel gas unit | FFGU |
| field pressure test flow diagram | FPTFD |
| field purchase order | FPO |
| field receiving report | FRR |
| field wildcat | FWC |
| figure | fig |
| fill up | FU |
| fillet weld | FW |
| filter | FLTR |
| filter cake | FC |
| filtrate | filt |
| final | fin |
| final (flowing) tubing pressure | FTP |
| final boiling point | FBP |
| final bottom-hole pressure, flowing | FBHPF |
| final bottom-hole pressure, shut-in | FBHPSI |
| Final Environmental Impact Statement | FEIS |
| final flowing pressure | FFP |
| final fluid level | FFL |
| final hydrostatic pressure | FHP |

## Definitions with Abbreviations

| | |
|---|---|
| final open | FO |
| final pressure | FP |
| final report for rig | FRR |
| final report for well | FRW |
| final shutin pressure | FSIP |
| finding of no significant impact | FONSI |
| fine | fn |
| fine grained | f-gr |
| finely | fnly |
| finely crystalline | f/xin |
| finish | FNSH |
| finish (ed) | fin |
| finished drilling | fin drlg |
| finished in hole | FIH |
| fire water | F/WTR |
| fire-resistant oil | F-R oil |
| fireproof | fprf |
| Firm Transport | FT |
| fiscal year ending | FYE |
| fishing | fish, fsg |
| fishing for | FF |
| fissile | fisl |
| fittings | ftg |
| fixed | fxd |
| fixed carbon | FC |
| fixture | fix |
| flaky | flk |
| flammable | FLMB |
| flammable liquid building | FL/BD |
| flange (ed) (es) | flg (d) (s) |
| flanged and dished (heads) | F&D |
| flange and screwed | F/S |
| flanged and spigot | F&S |
| flanged gate valve | FGVV |

| | |
|---|---|
| flanged one end, welded one end | FOE-WOE |
| flash point, Cleveland Open Cup | Fl-COC |
| flashing | FL |
| flat face | FF |
| flathead | Flath |
| flattened | flat |
| flexible | flex |
| flexibox | FLXBX |
| Flippen | Flip |
| float | flt |
| float collar | FC |
| float shoe | FS |
| flotation | fltn |
| floating | fltg |
| floating production storage and offloading vessel | FPSO |
| floor | FL |
| floor drain | FD |
| Florence flint | Flor fl |
| flow | flo |
| flow (ed) (ing) | flw (d) (g) |
| flow control valve | FCV |
| flow diagram | F-DIA, F/DIA, FD |
| flow indicating controller | FIC |
| flow indicating ratio controller | FIRC |
| flow indicator | FI |
| flow line | FL |
| flowmeter | F-MET |
| flow rate | FR |
| flow recorder | FR |
| flow recorder control | FRC |
| flow sheet | F-SHT |
| flow station | FS |
| flow switch | F/SW |
| flowed (ing) at the rate of | FARO |

## Definitions with Abbreviations

| | |
|---|---|
| flowed, flowing | Fl/, fl/ |
| flowerpot | Flwrpt |
| flowing | flg |
| flowing bottom-hole pressure | FBHP |
| flowing by heads | FBH |
| flowing casing pressure | FCP |
| flowing on test | FOT |
| flowing pressure | Flwg Pr, FP |
| flowing surface pressure | FSP |
| flue | flu |
| fluid | flu |
| fluid catalytic cracking | FCC |
| fluid in hole | FIH |
| fluid level | FL |
| fluid to surface | FTS |
| fluorescence, fluorescent | fluor |
| flush | FL |
| flush joint | FJ |
| flushed | flshd |
| flushing oil | FLO |
| focused log | FOCL |
| foliated | fol |
| foot-candle | ft-c |
| foot-pound | ft-lb |
| foot-pound per hour | ft-lb/hr |
| foot-pound-second (system) | f-p-s |
| footing, footage | ftg |
| for example | e.g. |
| for your information | FYI |
| Foraker | Forak |
| foraminifera | foram |
| foreman | f'man |
| forge (ed) (ing) | FRG |
| forged steel | FST, FS |

195

| | |
|---|---|
| formation | fm |
| formation density | F-D |
| formation density correlated | FDC |
| formation density log | FDL |
| formation gas-oil ratio | F/GOR |
| formation interval tester | FIT |
| formation test | FT |
| formation water | Fm W |
| Fort Chadborne | Ft C |
| Fort Hayes | Ft H |
| Fort Riley | Ft R |
| Fort Union | Ft U |
| Fort Worth | Ft W |
| Fortura | Fort |
| forward | fwd |
| fossiliferous | foss |
| foundation | fdn |
| Fountain | Fount |
| four pole single throw switch | 4P ST SW |
| four pole switch | 4P SW |
| four-wheel drive | FWD |
| Fox Hills | Fox H |
| frac finder (log) | FF |
| fractional | fr |
| fractionation, fractionator, fractional | fract |
| fracture gratient | F.G. |
| fracture, fractured, fractures | frac (d) (s) |
| fragment | frag |
| frame, framing | FRM (G) |
| framework | frwk |
| franchise | fran |
| Franconia | Franc |
| Fredericksburg | Fred |
| Fredonia | Fred |

## Definitions with Abbreviations

| | |
|---|---|
| free on board | FOB |
| free point back off | FPBO |
| free-point indicator | FPI |
| freezer | frzr |
| freezing point | FP |
| freight | frt |
| frequency | freq |
| frequency meter | FM |
| frequency modulation | FM |
| fresh | frs |
| fresh break | FB |
| fresh water | FW |
| fresh water | fwtr |
| friable | fri |
| friction reducing agent | FRA |
| froggy | Frgy |
| from | fr |
| from east line | FE/L, FEL, fr E/L |
| from north line | FN/L, FNL, fr N/L |
| from northeast line | FNEL |
| from northwest line | FNWL |
| from south and west lines | FS&WLs |
| from south line | FS/L, FSL, fr S/L |
| from southeast line | FSEL |
| from southwest line | FSWL |
| from west line | fr W/L, FWL, fr W/L |
| front | fr |
| front & side | F/S |
| frontier | Fron |
| frosted | fros, fr |
| frosted quartz grains | FQG |
| Fruitland | Fruit |
| fuel gas | FG |
| fuel oil | FO |

| | |
|---|---|
| fuel oil equivalent | F.O.E |
| fuel oil return | FOR |
| fuel oil supply | FOS |
| fuels & fabrication | F&F |
| fuels & lubricants | F&L |
| full freight allowed (purchasing term) | FFA |
| full hole | FH |
| full length drift | FLD |
| full of fluid | FF |
| full open head | FOH |
| full opening | FO |
| Fullerton | Full |
| functional check out | FCO |
| funnel viscosity | FV |
| furfural | furf |
| furnace fuel oil | FFO |
| furnace | furn |
| furnish (ed) | FURN |
| furniture and fixtures | furn & fix |
| Fuson | Fus |
| Fusselman | Fussel |
| Fusulinid | Fusul |
| future | fut |

## G

| | |
|---|---|
| gauge (ed) (ing) | ga |
| Galena | glna |
| Gallatin | Gall |
| galled threads | gld thd |
| gallon (s) | gal |
| gallons acid | GA |
| gallons breakdown acid | GBDA |

## Definitions with Abbreviations

| | |
|---|---|
| gallons condensate per day | GCPD |
| gallons condensate per hour | GCPH |
| gallons gelled water | GGW |
| gallons heavy oil | GHO |
| gallons mud acid | GMA |
| gallons of oil per day | GOPD |
| gallons of oil per hour | GOPH |
| gallons of solution | gal sol |
| gallons of water per hour | GWPH |
| gallons oil | GO |
| gallons per day | GPD |
| gallons per hour | GPH |
| gas-liquid ratio | GLR |
| gas odor | GO |
| gas odor distillate taste | GODT |
| gas pay | GP |
| Gas Production Unit | GPU |
| gas purchase contract | GPC |
| Gas Research Institute | GRI |
| gas reserve group | GRG |
| gas rock | G Rk |
| gas sales contract | GSC |
| gas show | GS |
| gas to surface | GRS |
| gas to surface (time) | GTS |
| gas too small to measure | GTSTM |
| gas unit | GU |
| gas volume | GV |
| gas volume not measured | GVNM |
| gas well | GW |
| gas well shut in | GSI |
| gas-fluid ratio | GFR |
| gas-handling study group | GHSG |
| gas-oil contact | GOC |

| | |
|---|---|
| gas-oil ratio | G/O, GOR |
| gas-water contact | GWC |
| gas-well gas | GWG |
| gaseous nitrogen | $G-N_2$ |
| gaseous oxygen | $G-O_2$ |
| gasket | gskt |
| gasoline | gaso |
| gasoline plant | GP |
| gastropod | gast |
| gathering line | G/L |
| gauge | gge |
| gauge (ed) (ing) | ga |
| gauge ring | GR |
| gelled | gel |
| general | genl |
| general agreement | GA |
| General Electric Company | GE |
| General Land Office (Texas) | GLO |
| General Motors Corporation | GM |
| generation, generator | gen |
| geological quadrangle map | GQM |
| geology (ist) (ical) | geol |
| geophysical investigation map | GPM |
| geophysics (ical) | geop |
| geopressure development, failure | PD |
| geopressure development, success | PDS |
| Georgetown | Geo |
| geothermal | geo, GT |
| geothermal development, failure | GD |
| geothermal development, success | GDS |
| geothermal wildcat, failure | GW |
| geothermal wildcat, success | GWS |
| Gibson | Gib |
| Gigajoule | GJ |

## Definitions with Abbreviations

| | |
|---|---|
| Gilcrease | Gilc |
| gilsonite | gil |
| glass, glassy | gls (y) |
| glauconite, glauconitic | glau |
| Glen Dean lime | GD li |
| Glen Rose | GR |
| Glenwood | Glen |
| globe valve | GLBVV |
| Globigerina | Glob |
| Glorieta | Glor |
| glycol | glyc |
| gneiss | gns |
| going in hole | GIH |
| Golconda lime | Gol |
| good | gd |
| good fluorescence | GFLU |
| good odor & taste | gd o&t |
| good show of gas | GSG |
| good show oil | GSO |
| good show oil and gas | GSO&G |
| good well shutin | GSI |
| Goodland | Gdld, Good L |
| Goodwin | Gdwn |
| goose egg | G egg |
| Gorham | Gor |
| Gouldbusk | Gouldb |
| government | govt |
| governor | gov |
| grade | gr |
| grading | grdg |
| grading location | grdg loc |
| gradiomanometer | GRAD |
| gradual, gradually | grad |
| grain | gr |

| | |
|---|---|
| grained (as in fine-grained) | gnd |
| grains per gallon | gg, GPG |
| gram | gm |
| gram molecular weight | g mole |
| gram calorie | g-cal |
| Graneros | Granos |
| granite point field | GPF |
| granite wash | gran w |
| granite, granular | gran |
| grant (of land) | grt |
| granular | grnlr |
| graptolite | grap |
| grating | grtg |
| gravel | gvl |
| gravel packet | GVLPK |
| gravitometer | grvt |
| gravity | grav, gr, GTY, GRVTY |
| gravityo API | gr API |
| gravity meter | G.M. |
| gray | gry |
| Gray sand | Gr Sd |
| Grayburg | Grayb |
| Grayson | Gray |
| greywacke | gywk |
| grease | gr |
| greasy | gsy |
| green | grn |
| Green River | Grn Riv |
| green royalty | gr roy |
| green shale | grn sh |
| Greenhorn | GH |
| grind out | GO |
| gritty | grty |
| grooved | grv |

## Definitions with Abbreviations

| | |
|---|---|
| grooved ends | GE |
| gross | grs |
| gross acre-feet | GAF |
| gross weight | gr wt |
| ground | gr, grnd |
| ground level | GL |
| ground measurement (elevation) | GM |
| group | GRP |
| guard log | GRDL |
| guidance continuance tool | GCT |
| guide shoe | GS |
| Gulf Research and Development Company | GR&DC |
| G. Riv | Gull River |
| gummy | gmy |
| gun barrel | GB |
| gun perforate | G/P |
| Gunsite | Guns |
| gusset | GUS |
| gypsiferous | gypy |
| gypsum | gyp |
| Gyroidina | Gyr. |
| Gyroidina scal | Gyr. sc. |

| | |
|---|---|
| Hackberry | Hackb |
| hackly | hky |
| hand hole | HH |
| hand-control valve | HCV |
| handle | hdl |
| handling | HNDLG |
| handwheel | HND/WHL |
| hanger | hgr |

203

| | |
|---|---|
| Haragan | Hara |
| harbor | hbr |
| hard | hd |
| hard lime | hd li |
| hard sand | hd sd |
| Hardinsburg sand (local) | Hburg |
| hardness | hdns |
| hardware | hdwe |
| Haskell | Hask |
| Haynesville | Haynes |
| hazardous | haz |
| head | hd |
| header | hdr |
| headquarters | HQ |
| heat exchanger | HX, HTX |
| heat tracing (ed) | HT |
| heat-treated alloy | HTA |
| heat-treated, heater treater | HT |
| heater | htr |
| heating and ventilating | H&V |
| heating oil | HO |
| heating ventilating and air conditioning | HVAC |
| heavily | hvly |
| heavily (highly) gas-cut salt water | HGCSW |
| heavily (highly) gas-cut water | HGCW |
| heavily (highly) oil-cut mud | HOCM |
| heavily (highly) oil-cut salt water | HOCSW |
| heavily (highly) oil-cut water | HOCW |
| heavily (highly) water-cut mud | HWCM |
| heavily (highly) gas-cut mud | HGCM |
| heavily (highly) oil- and gas-cut mud | HO&GCM |
| heavy | hvy |
| heavy coker gas oil | HCGO |
| heavy cycle oil | HCO |

## Definitions with Abbreviations

| | |
|---|---|
| heavy duty | HD |
| heavy fuel oil | HFO |
| heavy gas oil | HVGO |
| heavy hydrocrackate | HUX |
| heavy oil | HO |
| heavy reformate | HR |
| heavy steel drum | HSD |
| Heebner | Heeb |
| height | hgt |
| heirs | hrs |
| held by production | HBP |
| hematite | hem |
| Herington | Her |
| Hermosa | Herm |
| Hertz | Hz |
| heterostegina | het |
| hex head | HEX HD |
| hexagon (al) | hex |
| hexane | hex |
| Hickory | Hick |
| high detergent | HD |
| high gas-oil ratio | HGOR |
| high pressure | HP |
| high temperature | HT |
| high tension | HT |
| high viscosity | HV |
| high viscosity index | HVI |
| high voltage | H-VOLT |
| high volume lift | HVL |
| high-level shutdown | HLSD |
| high-pressure gas | HPG |
| high-pressure gauge | HPG |
| high-resolution dipmeter | HRD |
| high-temperature shutdown | HTSD |

| | |
|---|---|
| hily | highly |
| highway | hwy |
| Hilliard | Hill |
| hockleyensis | hock |
| Hogshooter | Hog |
| hold down | HLDN |
| hole full of oil | HFO |
| hole full of salt water | HFSW |
| hole full of sulfur water | HR Sul W |
| hole full of water | HFW |
| hole opener | HO, H. O. |
| holes per foot | HPF |
| Hollandberg | Holl |
| Home Creek | Home Cr |
| home office | HO |
| hook up | HU |
| hookwall packer | HWP |
| hooper | hop |
| Hoover | Hov |
| horizontal | horiz |
| horsepower | HP |
| horsepower-hour | hp-hr |
| Hospah | Hosp |
| hot dry rock development, failure | HD |
| hot dry rock development, successful | HDS |
| hot dry rock wildcat, failure | HW |
| hot dry rock wildcat, success | HWS |
| hot oil tar | HOT |
| hot-rolled steel | HRS |
| hour (s) | hr, HRS |
| house (ed) (ing) | HSE |
| house brand (regular grade of gasoline) | HB |
| Hoxbar | Hox |
| Humblei | Humb |

## Definitions with Abbreviations

| | |
|---|---|
| Humphreys | Hump |
| hundred weight | cwt |
| Hunton | Hun |
| hydraulic | HYD |
| hydraulic pump | HP |
| Hydril | HYD |
| Hydril Thread | HYDT |
| Hydril Type A joint | HYDA |
| Hydril Type CA joint | HYDCA |
| Hydril Type CS Joint | HYDCS |
| hydro test | HYDRO |
| hydrocarbon | HC |
| hydrocarbon drain system | HCDS |
| hydrocracker | H/C, HC |
| hydrogendelsulfurization | HDS |
| hydrofining | hfg |
| hydrogen | $H_2$ |
| hydrogen delsulfurization | HDS |
| hydrogen ion concentration | pH |
| hydrogen sulfite | $H_2S$ |
| hydrogenation | HYGN |
| Hydrologic Investigations Atlas | HIA |
| hydrostatic heat | HH |
| hydrostatic pressure | HP |
| hydrostatic test | HST |
| hydrotreater | hydtr |
| hygiene | Hyg |
| hyperbolic constant | CSCH |
| hyperbolic contangent | COTH |
| hypotenus | COSH |
| hypotenuse | HYPO |
| hydraulic horsepower | HHP |

# I

| | |
|---|---|
| identification sign | I.D. sign |
| identify (ier) (ication) | IDENT |
| Idiomorpha | Idio |
| igneous | ign |
| ignition | IGN |
| illuminator (s) | ILUM |
| imbedded | imdbed |
| immediate (ly) | immed |
| Imperial | Imp |
| Imperial gallon | Imp gal |
| impervious | imperv |
| impounding | IMP |
| impression block | IB |
| in hole | IH |
| imbedded | imdbed |
| incandescent | incd |
| inch (es) | in. |
| inch-pound | in.-lb |
| inches mercury | in. Hg |
| inches per second | in./sec |
| incinerator, incineration | INCIN |
| include (ed) (ing) | incl |
| inclusions | incls |
| income (er) (ing) | INCM |
| incorporated | Inc. |
| increase (ed) (ing) | incr |
| increment | incr |
| indicate (s) (tion) | indic |
| indicated horsepower | IHP |
| indicated horsepower hour | IHPHR |
| indistinct | indst |
| individual | indiv |

## Definitions with Abbreviations

| | |
|---|---|
| induction | ind |
| induction electrical survey | IES |
| indurated | indr |
| inflammable liquid | Inf. L |
| inflammable solid | Inf. S |
| inflow performance rate | IPR |
| information | info |
| infrared | IR |
| inhibitor | inhib |
| initial | init |
| initial air block | IAB |
| initial boiling point | IBP |
| initial bottom-hole pressure | IBHP |
| initial bottom-hole pressure, flowing | IBHPF |
| initial bottom-hole pressure, shut in | IBHPSI |
| initial flowing pressure | IFP |
| initial fluid level | IFL |
| initial hydrostatic pressure | IHP |
| initial mud weight | IMW |
| initial open | IO |
| initial participating area | IPA |
| initial potential | IP |
| initial pressure | IP |
| initial production | IP |
| initial production flowed (ing) | IPF |
| initial production gas lift | IPG |
| initial production on intermitter | IPI |
| initial production plunger lift | IPL |
| initial production pumping | IPP |
| initial production swabbing | IPS |
| initial shutin tubing pressure | ISITP |
| initial shutin pressure (DST) | ISIP |
| initial vapor pressure | IVP |
| injection gas | IG |

| | |
|---|---|
| injection gas-oil ratio | IGOR |
| injection index | II |
| injection pressure | Inj Pr |
| injection rate | IR |
| injection water | IW |
| injection well | IW |
| injection, injected | inj |
| inland | inl |
| inlet | inl |
| inlet gate valve/inlet ball valve | IGV/IBV |
| Inoceramus | Inoc |
| input/output | I/O |
| inquire, inquiry | INQ |
| inside diameter | ID |
| inside screw (valve) | IS |
| insoluble | insol |
| inspect (ed) (ing) (tion) | insp |
| install (ed) (ing) (tion) | inst |
| install (ing) pumping equipment | IPE |
| installation operation & maintenance | I-O-M |
| installation (s) | instl |
| installing (ed) pumping equipment | INPE |
| instantaneous | inst |
| instantaneous shutin pressure (frac) | ISIP |
| institute | inst |
| instrument, instrumentation | instr |
| insulate | insul |
| insulate, insulation | ins |
| insurance | ins |
| integral | INT |
| integral joint | IJ |
| integrator | intgr |
| intention to drill | ITD |
| interbedded | interbd |

| | |
|---|---|
| interconnecting flow diagram | I/CFD |
| interconnection (ing) | I/C, INTCON |
| intercooler | inclr, INCLR |
| intercrystalline | inter-xln |
| interest | int |
| intergranular | inter-gran, irgr |
| interior | int, INTR |
| interlaminated | inter-lam, inlam |
| intermediate manifolds | IMF |
| intermediate pressure | IP |
| internal | int, INTL |
| internal flush | IF |
| Internal Revenue Service | IRS |
| internal upset ends | IUE |
| Interruptible Transport | IT |
| intersect | ints |
| intersection | int |
| interstitial | intl |
| interval | intv |
| intrusion | intr |
| invert (ed) | inv |
| invertebrate | invrtb |
| investment tax credit | ITC |
| invitation to bid | ITB |
| invoice | inv |
| Ireton | Ire |
| iridescent | irid |
| iron | Fe |
| iron body (valve) | IB |
| iron body brass (bronze) mounted (valve) | IBBM |
| iron body brass core (valve) | IBBC |
| iron case | IC |
| iron pipe size | IPS |
| iron pipe thread | IPT |

| | |
|---|---|
| ironstone | Fe-st, irst |
| irregular | irreg |
| isocracker | ISO/CKR |
| isolate (tor) | ISOL |
| isometric | isom, ISO |
| isopropyl alcohol | IPA |
| isothermal | isoth |
| issue | ISS |
| Iverson | Ives |

## J

| | |
|---|---|
| jacket | jac |
| Jackson | Jxn, Jack |
| Jackson sand | Jax sd |
| jammed | jmd |
| jasper (oid) | Jasp |
| Jefferson | Jeff |
| jelly-like colloidal suspension | gel |
| jet fuel (aviation) | JFA |
| jet perforated | JP |
| jet perforations per foot | JP/ft |
| jet propulsion fuel | JP fuel |
| jet shots per foot | JSPF |
| jet treating unit | JTU |
| job complete | JC |
| jobber | jbr |
| Joint Committee on Uniformity of Methods of Water Examination | JCUMWE |
| Joint Interest Billing | JIB |
| joint interest nonoperated (property) | JINO |
| joint operating agreement | JOA |
| joint operating provisions | JOP |

| | |
|---|---|
| joint operation | J/O |
| joint venture | JV |
| joint (s) | jt(s) |
| Jordan | Jdn |
| Joule | J |
| jobber | jbr |
| Judith River | Jud Riv |
| junction | jct |
| junction box | JB |
| junk (ed) | jnk |
| junk basket | JB |
| junk joint | JJ |
| junked and abandoned | J&A |
| Jurassic | Jur |
| jurisdiction | juris |

| | |
|---|---|
| killed | kld |
| kiln dried | KD |
| kilocalorie | kcal |
| kilocycle | kc |
| kilogram | kg |
| kilogram-calorie | kg-cal |
| kilogram-meter | kg-m |
| kilohertz | kHz |
| kiloliter | kl |
| kilometer | km |
| kilopascal | kPa |
| kilovar-hour | kvar-hr |
| kilovar; reactive kilovolt-ampere | kvar |
| kilovolt | kv |
| kilovolt peak | kvp |

| | |
|---|---|
| kilovolt-ampere | kva |
| kilovolt-ampere-hour | kvah |
| kilowatt | kw |
| kilowatt-hour | kw-h |
| kilowatt-hour-meter | kw-hm |
| Kincaid lime | Kin, Li, KD |
| Kinderhook | Khk |
| kinematic | Kin |
| kinematic viscosity | KV |
| Kirtland | Kirt |
| kitchen | KIT |
| KMA sand | KMA |
| knock down | KD |
| knock out | KO |
| known geothermal resource area | KGRA |
| Kootenai | Koot |
| Krider | Kri |

## L

| | |
|---|---|
| La Motte | La Mte |
| labor | lab |
| laboratory | lab |
| ladder | lad |
| ladder & platform | L&P |
| lagging | LAG |
| laid (laying) down drill collars | LDDCs |
| laid (laying) down drillpipe | LDDP |
| laid down | LD |
| laid-down cost | LDC |
| Lakota | Lak |
| laminated, lamination (s) | lam |
| landing | LDG |

| | |
|---|---|
| land (s) | ld(s) |
| Landulina | Land |
| Lansing | Lans |
| lap joint | LJ |
| lapweld | LW |
| Laramie | Lar |
| large | lrg, lg |
| large-diameter flow line | LDF |
| Large Discorbis | Lg Disc |
| latitude | lat |
| Lauders | Laud |
| Layton | Layt |
| Le Comptom | Le C |
| leached | lchd |
| lead drill collar | LDCX |
| Leadville | Leadv |
| league | Lge |
| leak | lk |
| leakage | LKG |
| lease | lse |
| lease automatic custody transfer | LACT |
| lease crude | LC |
| lease operations | LOS |
| lease use (gas) | L U |
| Leavenworth | Lvnwth |
| left hand | LH |
| left in hole | LIH |
| legal subdivision (Canada) | LSD |
| length | lg |
| length overall | LOA |
| Lennep | Len |
| lense | lns |
| lenticular | len |
| less than truckload | LTL |

| | |
|---|---|
| less-than carload lot | LCL |
| letter | ltr |
| level | lvl |
| level alarm | LA |
| level control valve | LCV |
| level controller | LC |
| level glass | lg |
| level indicator | LI |
| level indicator controller | LIC |
| level recorder | LR |
| level recorder controller | LRC |
| license | lic |
| Liebuscella | Lieb |
| light | lt |
| light barrel | LB |
| light brown oil stain | LBOS |
| light coker gas oil | LCGO |
| light fuel oil | LFO |
| light hytrocrackate | LUX |
| light iron barrel | LIB |
| light iron grease barrel | LIGB |
| light steel drum | LSD |
| lighting | ltg |
| lightning arrester | LA |
| lignite, lignitic | lig |
| lime, limestone | Li, lm, ls |
| limy | lmy |
| limy shale | Lmy sh |
| limit, limonite | lim |
| limited | ltd |
| line | LN |
| line pipe | L.P. |
| line pressure | LP |
| line, as in E/L (east line) | /L |

| | |
|---|---|
| linear | lin |
| linear foot | lin ft |
| liner | lnr, lin |
| linguloid | lngl |
| liquefaction | liqftn |
| liquefied petroleum gas | LP-Gas, LPG |
| liquid | liq |
| liquid level controller | LLC |
| liquid level gauge | LLG |
| liquid oxygen | LOX |
| liquid penetrate examination | LPE |
| list of components | LST/COMPTS |
| liter | l |
| lithographic | litho |
| little | ltl |
| load | ld |
| load acid | LA |
| load oil | LO |
| load water | LW |
| loader | LDR |
| loading | LDG |
| loading dock | L-DK |
| local | LCL |
| local control panel | LCP |
| Local Distribution Company | LDC |
| local injections plants | LIP |
| local purchase order | LPO |
| located, location | loc |
| location abandoned | loc abnd |
| location graded | loc gr |
| lock | lk |
| locker | LKR |
| lodge pole | LP |
| log mean temperature difference | LMTD |

| | |
|---|---|
| log total depth | LTD |
| logarithm (common) | log |
| logarithm (natural) | ln |
| long | lg |
| long coupling | LC |
| long handle / round point | LH/RP |
| long radius | LR |
| long-range automation plan | LRAP |
| long-range automatic plan | LRP |
| long string | LS |
| long-term tubing test | LTT |
| long threads and coupling | LT&C |
| longitude (inal) | long |
| loop | lp |
| lost circulation | LC |
| lost circulation material | LCM |
| Lovell | Lov |
| Lovington | Lov |
| low pressure | LP |
| low viscosity index | LVI |
| low voltage | L-VOLT |
| low water loss | LWL |
| low-pressure separation | LPS |
| low-pressure separator | LP sep |
| low-temperature extraction unit | LTX unit |
| low-temperature separation unit | LTS Unit |
| low-temperature shutdown | LTSD |
| lower | lwr, low |
| Lower Albany | L/Alb |
| lower anhydrite stringer | LAS |
| lower casing | LC |
| Lower Cretaceous | L/Cret |
| lower explosive limit | LEL |
| Lower Glen Dean | LGD |

Definitions with Abbreviations

| | |
|---|---|
| Lower Menard | LMn |
| lower tubing | LT |
| Lower Tuscaloosa | L/Tus |
| Lower, i.e., L/Gallup | L/ |
| lube oil | LO |
| lubricate (ed) (ing) (tion) | lub |
| Lueders | Lued |
| lug cover type (5-gallon can) | LC |
| lug cover wit pour spout | LCP |
| lumber | lbr |
| lumpy | Lmpy |
| lustre | lstr |

| | |
|---|---|
| machine | mach |
| Mackhank | Mack |
| Madison | Mad |
| magnetic particle examination | MPE |
| magnetic, magnetometer | mag |
| magnetomotive force | mmf |
| main reaction furnace, "merf" | MRF |
| maintenance | maint |
| Manitoban | Manit |
| major, majority | maj |
| making | MKG |
| male and female (joint) | M&F |
| male pipe thread | MPT |
| male to female angle | MFA |
| malleable | mall |
| malleable iron | MI |
| management | m'gmt |
| manager | mgr |

219

| | |
|---|---|
| manhole | MH |
| manifold | MF, man |
| Manning | Mann |
| manual | man, MNL |
| manually operated | man op |
| manufacture (er) | MFR |
| manufactured | mfd |
| manufacturing | mfg |
| mapping subcommittee | MSC |
| Maquoketa | Maq |
| Marble Falls | Mbl Fls |
| marcaroni | MT |
| Marchand | March |
| marginal | marg |
| Marginulina | Marg. |
| Marginulina coco | Marg. coco. |
| Marginulina flat | Marg. fl. |
| Marginulina round | Marg. rd. |
| Marginulina texana | Marg. tex. |
| marine | mar |
| marine gasoline | margas |
| marine rig | MR |
| marine terminal | M/T |
| Marine Tuscaloosa | M. Tus |
| marine wholesale distributors | MWD |
| market demand factor | MDF |
| market( ing) | mkt |
| Markham | Mark |
| marking | MRK |
| marlstone | mrlst |
| marly | Mly |
| Marmaton | Marm |
| maroon | mar |
| Marshal (ling) | MARSH |

| | |
|---|---|
| masking plate | M/PLT |
| Massilina pratti | Mass pr. |
| massive | mass |
| massive anhydrite | MA |
| master | mstr |
| master circuit board | MCB |
| master load list | MLL |
| material | mat'l, mtl |
| material safety data sheets | MSDS |
| material take-off | MTO |
| mathematics | Math |
| matrix | Mtx |
| matter | Mat |
| maximum | max |
| maximum & final pressure | M&FP |
| maximum allowable operating pressure | MAOP |
| maximum allowable working pressure | MAWP |
| maximum casing pressure | MCP |
| maximum daily delivery obligation | MDDO |
| maximum efficient rate | MER |
| maximum flowing pressure | MFP |
| maximum operating pressure | MOP |
| maximum pressure | MP |
| maximum surface pressure | MSP |
| maximum top pressure | MTP |
| maximum total depth | MTD |
| maximum tubing pressure | MTP |
| maximum working pressure | MWP |
| Maywood | May |
| McClosky lime | McC lm |
| McCullough | McCul |
| McElroy | McEl |
| McKee | McK |
| McLish | McL |

| | |
|---|---|
| McMillan | McMill |
| Meakin | Meak |
| mean effective pressure | MEP |
| mean low water to platform | MLW-PLAT |
| mean low wave | MLW |
| mean sea level | msl |
| mean temperature difference | MTD |
| measure (ed) (ment) | meas |
| measure of hydrogen potential | pH |
| measured depth | MD |
| measured total depth | MTD |
| measuring and regulating station | M&R Sta. |
| mechanic (al), mechanism | mech |
| mechanical down time | Mech DT |
| mechanical flow diagram | MFD |
| mechanical properties | MPL |
| mechanism | mchsm |
| median | med |
| Medicine Bow | Med B |
| Medina | Med |
| medium | med |
| medium amber cut | MAC |
| medium fuel oil | med FO |
| medium grained | m-gr, med gr, mg |
| Medrano | Medr |
| Meeteetse | Meet |
| megacycle | mc |
| megahertz (megacycles per second) | MHz |
| megapascal | mPa |
| megawatt | MW |
| melting point | MP |
| member (geologic) | mbr |
| memo requesting quote (s) | MRQ |
| memorandum | memo |

| | |
|---|---|
| Menard lime | Men li |
| Menefee | Mene |
| Meramec | Mer |
| mercaptan | mercap |
| merchandise | mdse |
| mercury | merc |
| "merf", main reaction surface | MRF |
| meridian | merid |
| Mesaverde | Mvde |
| mesh | M |
| Mesozoic | Meso |
| metal petal basket | MPB |
| metamorphic | meta |
| meter | m, mtr |
| meter run | MR |
| meter-kilogram | m-kg |
| methanator | methr |
| methane | meth |
| methane-rich gas | MEG |
| methanol | methol |
| methyl chloride | meth-cl |
| methylene blue | meth-bl |
| methylethylketone | MEK |
| methylisobutylketone | MIK |
| metric | metr |
| mezzanine | mezz |
| mhos per meter | mho/m |
| mica, micaceous | mica |
| micro-microfarad | m-µf |
| microampere | µA |
| microcrystalline | micro-xn, micro-x |
| microfarad | µfd, m-f |
| microfossil (iferous) | micfos |
| microgram | µ-g |

| | |
|---|---|
| microgram (millionith) | µg |
| microinch | µ-in. |
| micromicron | µ-µ |
| micron | µ |
| microsecond | microsec., µsec |
| microvolt | µv |
| microwave | MW |
| middle | M/, mdl, mid |
| middle ground shoals | MGS |
| middle hydrocrackate | MUX |
| Midway | Mwy |
| mile (s) | mi |
| miles per hour | MPH |
| miles per year | MPY |
| military | mil |
| milky | mky |
| mill wrapped plain end | MWPE |
| milled | mld |
| milled one end | Mle |
| milled one end | M1E |
| milled other end | MOE |
| milled two ends | M2E |
| milliampere | ma |
| millidarcies | md |
| milligram | mg |
| millihenry | mh |
| milliliter | ml |
| milliliters tetraethyl lead per gallon | ml TEL/G |
| millimeter | mm |
| millimeters of mercury | mm Hg |
| millimicron | mµ |
| milling | millg |
| milling | mlg |
| million (i.e., 9MM = 9,000,000) | MM |

## Definitions with Abbreviations

| | |
|---|---|
| million British thermal units | MMBTU |
| million cubic feet | MMcf |
| million cubic feet per day | MMcfd |
| million electron volts | MMev |
| million reservoir barrels | MMRVB |
| million standard cubic feet per day | MMscfd |
| millions of barrels | MMBLS |
| milliotitic | mill |
| milliroentgen | mr |
| millisecond (s) | ms |
| millivolt | mv |
| mineral | mnrl |
| mineral interest | MI |
| minerals | min |
| Minerals Management Service | MMS |
| minimum | min |
| Minimum Daily Quantity | MDQ |
| minimum pressure | min P |
| Minnekahta | Mkta |
| Minnelusa | Minl |
| minor | MNR |
| minute (s) | min |
| Miocene | Mio |
| miscellaneous | misc |
| Miscellaneous Field Studies Map | MFSM |
| miscible substance group | MSWG |
| Misener | Mise |
| Mission Canyon | Miss Cany |
| Mississippian | Miss |
| mix and pump | M&P |
| mixed | mxd |
| mixer | mix |
| mixing | MIXG |
| mobile | mob |

| | |
|---|---|
| model | mod |
| moderate (ly) | mod |
| modification | mod |
| modular | modu |
| Moenkopi | Moen |
| moisture impurities and unsaponifiabales (grease-testing) | MIU |
| molar | M |
| molas | mol |
| mole | mol |
| molecular weight | mol wt |
| molecule, molecular | MOL |
| mollusca | mol |
| molybdenum | MO |
| Monel drill collars | MDC |
| monitor | mon |
| monoethanolamine | MEA |
| monthly volume operation plan | MVOP |
| Montoya | Mont |
| Moody's Branch | MB |
| Moore County lime | MC ls |
| Mooringsport | Moor |
| more or less | m/l |
| Morrison | Morr |
| Morrow | Mor |
| mortgage | mtge |
| Mosby | Mos |
| motor | MTR, mot |
| motor control center | MCC |
| motor generator | mg |
| motor medium | MM |
| motor octane number | MON |
| motor oil | MO |
| motor oil units | MOU |

## Definitions with Abbreviations

| | |
|---|---|
| motor severe | MS |
| motor vehicle fuel tax | MVFT |
| motor vehicle, motor vessel | M/V |
| mottled | mott |
| Mount Selman | Mt. Selm |
| mounted | mtd |
| mounting | mtg |
| mousehole | MH |
| moving | movg |
| moving (moved) in double drum unit | MIDDU |
| moving in (equipment) | MI |
| moving in and rigging up | MIRU |
| moving in cable tools | MICT |
| moving in completion rig | MICR |
| moving in equipment | MIE |
| moving in materials | MIM |
| moving in pulling unit | MIPU |
| moving in rig | MIR |
| moving in rigging up swabbing units | MIRUSU |
| moving in rotary tools | MIRT |
| moving in service rig | MISR |
| moving in standard tools | MIST |
| moving in tools | MIT |
| moving out | MO |
| moving out (off) cable tools | MOCT |
| moving out (off) rotary tools | MORT |
| moving out completion unit | MOCU |
| moving out equipment | MOE |
| moving out rig | MOR |
| Mowry | Mow |
| Mt. Diablo | MD |
| mud acid | MA |
| mud acid wash | MAW |
| mud cake | MC |

227

| | |
|---|---|
| mud cleanout agent | MCA |
| mud cut | MC |
| mud filtrate | MF |
| mud logger | ML |
| mud logging unit | MLU |
| mud to surface | MTS |
| mud weight | md wt, MW |
| mud-cut acid | MCA |
| mud-cut gas | MCG |
| mud-cut oil | MCO |
| mud-cut salt water | MCSW |
| mud-cut wayer | MCW |
| mud/silt remover | MSR |
| muddy | Mdy |
| muddy salt water | MSW |
| muddy water | MW |
| mudstone | mudst |
| multigrade | MG |
| multiple service acid | MSA |
| multiply, multiplexer | MULTX |
| multipurpose | MP |
| multipurpose grease lithium base | MPGH-lith |
| multipurpose grease soap base | MPGR-soap |
| muscovite | musc |

## N

| | |
|---|---|
| Nacotoch | Nac |
| nacreous | nac |
| nameplate | NP |
| naphfining unit | NU |
| naptha | nap |
| naptha-hydrogen desulfurization | NHDS |
| narrative | NARR |

## Definitions with Abbreviations

| | |
|---|---|
| national | nat'l |
| national coarse thread | NCT |
| National Electric Code | NEC |
| National Fine (thread) | NF |
| National pipe thread | NPT |
| National pipe thread, female | NPTF |
| National pipe thread, male | NPTM |
| National pollution discharge elimination system | NPDES |
| natural | nat |
| natural flow | NF |
| natural gas | NG |
| natural gas liquids | NGL |
| nautical line | NMI |
| Navajo | Nav |
| Navarro | Navr |
| Naval Petroleum Reserve | NPR |
| Naval Petroleum Reserve, Alaska | NPRA |
| negative | neg |
| negligible | neg |
| negotiation | NEGO |
| neoprene | npne |
| net effective pay | NEP |
| net positive suction head | NPSH |
| net revenue interest | NRI |
| net tons | NT |
| neutral neutralization | neut |
| neutralization number | Nuet. No. |
| neutron lifetime log | NLL |
| New Albany shale | New Alb |
| new bit | NB |
| new field discovery | NFD |
| new field wildcat | NFW |
| new field wildcat, discovery | WFD |
| new field wildcat, dry new oil | WF |

229

| | |
|---|---|
| new oil | NO |
| new pool discovery | NPD |
| new pool exempt (nonprorated) | NPX |
| new pool wildcat | NPW |
| new pool wildcat, discovery | WPD |
| new pool wildcat, dry new rod | WP |
| new rod | NR |
| new source performance standards | NSPS |
| new total depth | NTD |
| Newburg | Nbg |
| Newcastle | Newc |
| Niagra | Nig |
| nickel plated | NP |
| Ninnescah | Nine |
| Niobrara | Niob |
| nipple | nip, NPL |
| nipple (ed) (ing) down blowout preventers | NDBOPs |
| nipple (ed) (ing) up blowout prevents | NUBOPs |
| nipple-down tree | NDT |
| nipple-up tree | NUT |
| nippled (ing) up | NU, UP |
| nippled down | ND |
| nippling up wellhead | NUWH |
| nitrogen | $N_2$ |
| nitrogen blanket | NB |
| nitroglycerine | nitro |
| no appreciable gas | NAG |
| no change | NC |
| no core | NC |
| no fluid | NF |
| no fluorescence | NF |
| no fluorescence or cut | NFOC |
| no fuel | NF |
| no gas to surface | NGTS |

| | |
|---|---|
| no gauge | NG |
| no good | NG |
| no increase | No Inc |
| no order required | NOR |
| no paint on seams | NPOS |
| no production | NP |
| no recovery | no rec, NR |
| no report, not reported | NR |
| no show | NS |
| no show fluorescence or cut | NSFOC |
| no show gas | NSG |
| no show oil | NSO |
| no test | N/tst |
| no time | NT |
| no visible porosity | NVP |
| no water | NW |
| Noble-Olson | NO |
| Nodosaria blanpiedi | Nod. blan. |
| Nodosaria mexicana | Nod. Mex. |
| nodule, nodular | nod |
| nominal | nom |
| nominal pipe size | NPS |
| noncontiguous tract | NCT |
| nondestructive testing | NDT |
| nondetergent | ND |
| nonemulsifying agent | NE |
| nonemulsion acid | NEA |
| nonleaded gas | NL Gas |
| nonoperated joint ventures | NOJV |
| nonoperating property | NOP |
| nonporous | NP |
| nonproducer | NP |
| nonreturnable steel barrel | NRSB |
| nonreturnable steel drum | NRSD |

| | |
|---|---|
| nonreturnable, no returns, not reached | NR |
| nonrising stem (valve) | NRS |
| nonstandard | nstd |
| nonstandard service station | N/S S/S |
| nonupset | NU |
| nonupset ends | NUE |
| nonemulsifying agent | NE |
| nonflammable compressed gas | nonf G |
| Nonionella | Non |
| Nonionella Cockfieldensis | N. Cock. |
| Noodle Creek | Ndl Cr |
| normal | nor |
| normal (to express concentration) | N |
| normally closed | NC |
| normally open | NO |
| north | N |
| north half | N/2 |
| north line | NL |
| north offset | N/O |
| north quarter | N/4 |
| northeast | NE |
| northeast corner | NEC |
| northeast line | NEL |
| northeast quarter | NE/4 |
| northerly | N'ly |
| Northern Alberta Land Registration | NALRD |
| northwest | NW |
| northwest corner | NW/C |
| northwest line | NWL |
| northwest quarter | NW/4 |
| Northwest Territories | NWT |
| not applicable | NA |
| not available | NA |
| not completed | NC |

## Definitions with Abbreviations

| | |
|---|---|
| not deep enough | NDE |
| not drilling | ND |
| not in contract | NIC |
| not on bottom | NOB |
| not prorated | NP |
| not pumping | NP |
| not suitable for coating | NSC |
| not to scale | NTS |
| not yet available | NYA |
| not yet drilled | NYD |
| Notary Public | NP |
| notice of intention to drill | NID |
| notice of violation | NOV |
| notice to proceed | NTP |
| nozzle | noz |
| Nuclear Regulatory Commission | NRC |
| nugget | Nug |
| number | NO |
| number (before a number, i.e., No. 3) | No. |
| numerous | num |
| nylon | NYL |

## O

| | |
|---|---|
| Oakville | Oakv |
| object | obj |
| observation | OBS |
| obsolete | obsol |
| occasional (ly) | occ |
| ocean bottom suspension | OBS |
| octagon, octagonal | oct |
| octane | oct |
| octane number requirement | ONR |

| | |
|---|---|
| octane number requirement increase | ONRI |
| O'Dell | Odel |
| odor | od |
| odor, stain and fluorescence | OS&F |
| odor, taste & stain | OT&S |
| odor, taste, stain and fluorescence | OTS&F |
| off bottom | OB |
| offshore | offsh |
| off / on location | OL |
| office, official | off |
| official potential test | OPT |
| offsite | OFS |
| Ohio River Valley Water Sanitation Committee | ORSANCO |
| ohm | ohm |
| ohm-centimeter | ohm-cm |
| ohmmeter | ohm-m |
| oil | O |
| oil abandoned well | OAW |
| oil and gas | O&G |
| Oil and Gas Investigations Chart | OC- |
| Oil & Gas Journal | OGJ |
| oil and gas lease | O&GL |
| oil- and gas-cut acid water | O&GCAW |
| oil- and gas-cut load water | O&GCLW |
| oil- and gas-cut mud | O&GCM |
| oil- and gas-cut salt water | O&GCSW |
| oil- and gas-cut sulfur water | O&GC SULW |
| oil- and gas-cut water | O&GCW |
| oil and salt water | O&SW |
| oil and sulfur water-cut mud | O&SWCM |
| oil and water | O&W |
| oil-based mud | OBM |
| oil circuit breaker | OCB |
| Oil Creek | Oil Cr |

## Definitions with Abbreviations

| | |
|---|---|
| oil cut | OC |
| oil down to | ODT |
| oil emulsion | OE |
| oil emulsion mud | OEM |
| oil fluorescence | OFLU |
| oil fractured | oilfract |
| oil immersed, water cooled | OIWC |
| oil in hole | OIH |
| oil in place | OIP |
| oil in tanks | OIT |
| oil insulated | OI |
| oil insulated fan cooled | OFIC |
| oil insulated, self-cooled | OISC |
| oil odor | OO |
| oil pay | OP |
| oil payment interest | OPI |
| Oil Refining Industry Action Committee | ORIAC |
| oil sand | O sd |
| oil show | OS |
| oil stain | OSTN |
| oil standing in drillpipe | OSIDP |
| oil string flange | OSF |
| oil to surface | OTS |
| oil unit | OU |
| oil well flowing | OWF |
| oil well from water flood | OWFWF |
| oil well gas | OWG |
| oil well shut in | OSI |
| oil-cut mud | OCM |
| oil-cut salt water | OCSW |
| oil-cut water | OCW |
| oil-powered total energy | OTE |
| oil-soluble acid | OSA |
| oil-water contact | OWC |

| | |
|---|---|
| old plugback | OPB |
| old plugback depth | OPBD |
| old total depth | OTD |
| oil well drilled deeper | OWDD |
| old well plugged back | OWPD |
| old well sidetracked | OWST |
| old well worked over | OWWO |
| olefin | ole |
| Oligocene | Olig |
| on center | OC |
| one thousand foot-pounds | kip-ft |
| one thousand pounds | kip |
| ooliclastic | ooc |
| oolimoldic | oom |
| oolitic | ool |
| open (ed) (ing) | opn |
| open choke | OC |
| open cup | OC |
| open end | OE |
| open flow | OF |
| open flow potential | OFP |
| open hearth | OH |
| open hole | OH, op hole |
| open line (no choke) | OL |
| open-top tank | O/T tk |
| open tubing | OT |
| open-file report | OF |
| operate, operations, operator | oper |
| operation shutdown | OSD |
| operations and maintenance | O&M |
| operations commenced | OC |
| Operculinoides | Operc |
| opposite | opp |
| optimum bit weight and rotary speed | OBW&RS |

## Definitions with Abbreviations

| | |
|---|---|
| option to farmout | optn to F/O |
| optional | OPTL |
| orange | OR |
| Ordovican | Ord |
| Oread | Or |
| organic | org |
| organization | org |
| orientation | ORIENT |
| oriented microresistivity | OMRL |
| orifice | orf |
| orifice flange one end | OFOE |
| original oil in place | OOIP |
| original stock tank oil in place | OSTOIP |
| original total depth | OTD |
| original, originally | orig |
| Oriskany | Orisk |
| orthoclase | orth |
| Osage | Os |
| Osborne | O |
| Ostracod | ost |
| Oswego | Osw |
| other end beveled | OEB |
| ounce | oz |
| Ouray | Our |
| out of service over and short (report) | O/S |
| out of stock | O/S |
| outboard motor oil | OBMO |
| Outer Continental Shelf | OCS |
| outlet | otl, OUT |
| outline | OLN |
| outpost | OP |
| outside diameter | OD |
| outside screw and yoke (valve) | OS&Y |
| overexpenditure | OE |

| | |
|---|---|
| overhead | OH |
| overproduced | OP |
| overall | OA |
| overall height | OAH |
| overall length | OAL |
| overflush (ed) | OFL |
| overhead | ovhd |
| overriding royalty | ORR |
| overriding royalty interest | ORRI |
| overseas procurement office | OPO |
| overshot | OS |
| overtime | OT |
| oxidized, oxidation | ox |
| oxygen | oxy |

## P

| | |
|---|---|
| Pacific Outer Continental Shelf | POCS |
| packed | pkd |
| packer | pkr |
| packer set at | PSA |
| packing, package (ed) | pkg (d) |
| Paddock | Padd |
| page (before a number, i.e., p. 4) | p. |
| Pahasapa | Paha |
| paid | pd |
| Paint Creek | PC |
| pair | pr |
| paleontology | Paleo |
| Paleozoic | Paleo |
| Palo Pinto | Palo P |
| Paluxy | Pal, Pxy |
| panel | pnl |

## Definitions with Abbreviations

| | |
|---|---|
| panel board | PNL BD |
| Panhandle Lime | Pan L |
| Paradox | Para |
| Park City | Park C |
| parish | Ph |
| parrafins-olefins-napthenes-aromatics | PONA |
| party, partly | pt |
| partial | PART |
| participating area | PA |
| partings | prtgs |
| partition | PTN |
| partly | ptly |
| parts per billion | ppb |
| parts per million | ppm |
| Pascal | Pa |
| patent (ed) | pat |
| pattern | patn |
| pavement | pvmt |
| paving | pav |
| Pawhuska | Paw |
| payment | pymt |
| pearly | prly |
| pebble, pebbly | pbl (y) |
| Pecan Gap | PG |
| Pecan Gap chalk | PGC |
| pedestal | PED |
| pelecypod | plcy |
| pellletal, pelletoidal | pell |
| penalty, penalize (ed) (ing) | penal |
| penetration asphalt cement | Pen A.C. |
| penetration index | PI |
| penetration, penetration test | pen |
| Pennsylvanian | Penn |
| Pensky Martins | P-M |

| | |
|---|---|
| Pensky-Martins (flash) | P-M |
| per-acre bonus | PAB |
| per-acre rental | PAR |
| percent | pct, % |
| per day | PD |
| per foot | /ft, pft |
| percolation | perco |
| perforate (ed) (ing) (or) | perf |
| perforated casing | perf csg |
| perforating hyper select | H-SEL |
| perforating, Enerjet | EJ |
| perforating, Hyperdome | HD |
| perforating, Hyperdome II | P-HDII |
| perforating, Ultrajet | ULJ |
| performance evaluation and review technique | PERT |
| Performance Number (aviation gas) | PN |
| period | prd |
| peripheral wedge zone | PWZ |
| permanent | perm |
| permanent type completion | PTC |
| permanently shut down | PSD |
| permeable (ability) | perm |
| Permian | Perm |
| permit | prmt |
| perpendicular | perp |
| personal and confidential | P&C |
| personnel | pers |
| petrochemical | petrochem |
| petroleum | pet |
| petroleum and natural gas | P&NG |
| petroliferous | petrf |
| Pettet | Pet |
| Pettit | Pett |
| Pettus sand | Pet sd |

| | |
|---|---|
| phase | ph |
| Phosphoria | Phos |
| Phrohotite | po |
| picked up | PU |
| picking up drillpipe | PUDP |
| Pictured Cliff | Pic Cl |
| piece | pc |
| pilot | plt |
| pilot loaded valve | PLV |
| pin end | pe |
| Pin Oak | P.O. |
| Pine Island | PI |
| pink | pk |
| pinpoint | pinpt, PP |
| pinpoint porosity | PPP |
| pint | pt |
| pipe buttweld | PBW |
| pipe electric weld | PEW |
| pipe lapweld | PLW |
| pipe seamless | PSM |
| pipe sleeve | PSL |
| pipe spiral weld | PSW |
| pipe to soil potential | PTS pot |
| pipe-handling capacity | PHC |
| pipeline | PL |
| pipeline oil | PLO |
| pipeline terminal | PLT |
| pipeway | PWY |
| piping | ppg |
| piping and instrument diagrams | P&IDS |
| piping diagram | P/DIA |
| pisolites, pisolitic | piso |
| pitted | pit |
| plagioclase | plg |

| | |
|---|---|
| plain both ends | PBE |
| plain end | PE |
| plain end beveled | PEB |
| plain large end | PLE |
| plain one end | POE |
| plain small end | PSE |
| plan | pln |
| plan of development | POD |
| plant | plt |
| plant (pressure) volume reduction | PVR |
| plant fossils | pl fos |
| Planulina harangensis | Plan. har. |
| Planulina palmarie | Plan. palm. |
| plaster | PLASR |
| plastic | plas |
| plastic viscosity | PV |
| plate | PL |
| platform | platf |
| platy | plty |
| please note and return | PNR |
| Pleistocene | Pleist |
| Pliocene | Plio |
| plug | Pg |
| plug down | PD |
| plug on bottom | POB |
| plugged | plgd |
| plugged and abandoned | P&A |
| plugged back | PB |
| plugged-back depth | PBD |
| plugged back total depth | PBTD |
| plumbing | PLMB |
| plunger | plngr |
| pneumatic | pneu |
| Podbielniak | Pod. |

| | |
|---|---|
| point | pt |
| Point Lookout | Pt Lkt |
| poison | pois |
| poker chipped | PC |
| polish (ed) | pol |
| polished rod | PR |
| polyethylene | poly cl |
| polymerization, polymerized | poly |
| polymerized gasoline | polygas |
| polypropylene | polypl |
| polyvinyl chloride | PVC |
| Pontotoc | Pont |
| pooling agreement | PA |
| poor | pr |
| porcelaneous | porc |
| porcion | porc |
| pore volume | PV |
| porosity and permeability | P&P |
| porosity, porous | por |
| porous and permeable | P&P |
| port collar | pc |
| portable | port |
| Porter Creek | PC |
| position | pos |
| positive | pos |
| positive crankcase ventilation | PCV |
| possible (ly) | poss |
| Post Laramine | P Lar |
| Post Oak | P.O. |
| postweld heat treatment | PWHT |
| potable water | POT/WTR |
| potential | pot |
| potential difference | pot dif |
| potential test | PT |

| | |
|---|---|
| potential test to follow | PTTF |
| pound | lb |
| pound-inch | lb-in. |
| pounds per barrel | PPB |
| pounds per cubic foot | PCF |
| pounds per gallon | PPG |
| pounds per square foot | psf, lb/sq ft |
| pounds per square inch | psi, PSI |
| pounds per square inch absolute | psia, PSIA |
| pounds per square inch gauge | psig, PSIG |
| pour point (ASTM method) | pour ASTM |
| power | PWR |
| power distribution center | PDC |
| power distribution system | PDS |
| power factor | PF |
| power factor meter | PFM |
| Precambrian | Pre Camb |
| precast | prcst |
| precipitate | ppt |
| precipitation number | ppn no |
| precipitator | PRECIP |
| predominant | predom |
| prefabricated | prefab |
| preferred | pfd |
| prefractionator | PFRACT |
| preheater | prehtr |
| preliminary | prelim |
| premium | prem |
| prepaid | ppd |
| prepare, preparing, preparation | Prep |
| preparing to take potential test | PRPT |
| present depth | PD |
| present operations | pr op |
| present production | P.P. |

## Definitions with Abbreviations

| | |
|---|---|
| present total depth | PTD |
| present worth at discount rate of 15% | PW(15) |
| pressed distillate | PD |
| pressure | press |
| pressure alarm | PA |
| Pressure Base | PB |
| pressure control valve | PCV |
| pressure differential controller | PDC |
| pressure differential indicator | PDI |
| pressure differential indicator controller | PDIC |
| pressure differential recorder | PDR |
| pressure differential recorder controller | PDRC |
| pressure indicator | PI |
| pressure indicator controller | PIC |
| pressure recorder | PR |
| pressure recorder control | PRC |
| pressure safety valve | PSV |
| pressure seal bonnet | PSB |
| pressure switch | PS |
| pressure-volume temperature | PVT |
| prestressed | prest |
| prevent, preventive | prev |
| prevention of significant deterioration, EPA | PSD |
| previous | PREV |
| previous daily average | PREV DO AVG |
| primary | pri |
| Primary Reference Fuel | PRF |
| principal | prin |
| principal lessee (s) | prncpl lss |
| prism (atic) | pris |
| private branch exchange | PBX |
| privilege | priv |
| probable (ly) | prob |
| process | proc |

| | |
|---|---|
| process & instrument diagram | P&ID |
| process flow diagram | PFD |
| Process Performance Index | PPI |
| produce (ed) (ing) (tion), product (s) | prod |
| producing gas well | PGW |
| producing oil and gas well | POGW |
| producing oil well | POW |
| producing oil well, flowing | POWF |
| producing oil well, pumping | POWP |
| producing well | PW |
| production department exploratory test | PDET |
| production management | PML |
| production payment | PP |
| production payment interest | PPI |
| production test flowed | PTF |
| production test pumped | PTP |
| productivity index | PI |
| professional paper | PP |
| profit and loss | P&L |
| profit-sharing interest | PSI |
| progress | prog |
| project (ed) (ion) | proj |
| project ultimate cost | PUC |
| projected total depth | PTD |
| propane | LPG |
| property | PROP |
| property line | PL |
| proportional | prop |
| propose (ed) | prop |
| proposed bottom-hole location | PBHL |
| proposed depth | PD |
| proposed total depth | PTD |
| prorated | pro |
| prospect | Psp |

## Definitions with Abbreviations

| | |
|---|---|
| protection | prot |
| Proterozoic | Protero |
| provincial | Prov |
| pseudo | pdso, ps |
| public address | PA |
| public relations | PR |
| Public School Land | PSL |
| pull (ed) rods and tubing | PR&T |
| pull (put) out of hole | POOH |
| pulled | pld |
| pulled (put) out of hole | POH |
| pulled bid pipe | PBP |
| pulled out | PO |
| pulled pipe | PP |
| pulled up | PU |
| pulled up in casing | PUIC |
| pulling | plg |
| pulling tubing | PTG |
| pulling tubing and rods | PTR |
| pulsation dampner | PD |
| pulse (sating) (sation) | PULS |
| pump | P/ |
| pump and flow | P&F |
| pump building | P/BLDG |
| pump in | PI |
| pump jack | PJ |
| pump job | PJ |
| pump on beam | POB |
| pump pressure | PP |
| pump station | PS |
| pump testing | P tstg |
| pump-in pressure | PIP |
| pump (ed) (ing) | pmp (d) (g) |
| pumper's depth | PD |

| | |
|---|---|
| pumping equipment | PE |
| pumping for test | PFT |
| pumping load oil pilot | PLO |
| pumping unit | PU |
| pumps off | PO |
| purchase order | PO |
| purchasing | PURCH |
| purchasing request | PR |
| purification | PURF |
| purple | purp |
| putting on pump | POP |
| pyrite, pyritic | pyr |
| pyrobitumen | pyrbit |
| pyroclastic | pyrclas |
| pyrolysis | pyls |

## Q

| | |
|---|---|
| quadrant | QDRNT |
| quadrant, quadrangle, quadruple | quad |
| qualitative | QUAL |
| quality | qual |
| quality assurance | QA |
| quality control | Q.C., QC |
| quality discount allowance | QDA |
| quantity | qty |
| quarry | qry |
| quart (s) | qt |
| quarter | qtr |
| quartz, quartzite, quartzitic | qtz |
| quartzose | qtzose |
| Queen City | Q. City |
| Queen Sand | Q. sd |

## Definitions with Abbreviations

| | |
|---|---|
| quench | qnch |
| questionable | quest |
| quick ram change | QRC |
| quintuplicate | quint |

## R

| | |
|---|---|
| rack | RK |
| radiant | RADT |
| radiation | radtn |
| radical | rad |
| radioactive | RA |
| radiographic examination | RT |
| radiological | rad |
| radius | R |
| radius | rad |
| railing | rlg |
| railroad | RR |
| Railroad Commission (Texas) | RRC |
| rainbow show of oil | RBSO |
| raised face | RF |
| raised face, flanged end | RFFE |
| raised face, slip on | RFSO |
| raised face, smooth finish | RFSF |
| raised face, weld neck | RFWN |
| Ramsbottom Carbon Residue | RCR |
| ran (running) rods and tubing | RR&T |
| ran in hole | RIH |
| random lengths | RL |
| range | rge |
| Ranger | Rang |
| rankine (temp.scale) | R |
| rapid curing | RC |

| | |
|---|---|
| rat hole | RH |
| rat hole mud | RHM |
| rate of penetration | ROP |
| rate of return | ROR |
| rate too low to measure | RTLTM |
| rating | rtg |
| raw gas | RG |
| raw gas lift | RAGL |
| re-evaluation for overoptimism | REFOO |
| reabsorber | REABS |
| reacidize (ed) (ing) | reacd |
| reaction (ed) | react |
| ready for rig | RFR |
| ream | rm |
| reamed | rmd |
| reaming | rmg |
| reboiler | RBLR |
| received | recd |
| receiver | recr |
| receptable | recp, RCPT |
| reciprocate (ing) | recip |
| recirculate | recirc |
| recommend | rec |
| recommended spare part | R-SP |
| recomplete (ed) (ion) | recomp |
| recompressor | RECOMP |
| recondition (ed) | recond |
| record (er) (ing) | rec |
| recover (ed) (ing), recovery | rec |
| recovery | RCVY |
| rectangle, rectangular | rect |
| rectifier | rect |
| recycle | recy, RCYL |
| recycle | recy, RCYL |

| | |
|---|---|
| red beds | Rd Bds |
| Red Cave | RC |
| Red Fork | Rd Fk |
| red indicating lamp | RIL |
| Red Oak | R.O. |
| Red Peak | Rd Fk |
| Red River | RR |
| redrilled | redrld, RR |
| reducer | RDCR |
| reducing balance | red bal |
| reducing, reducer | red |
| reference | ref |
| refine (ed) (er) (ry) | ref |
| Refinery Technology Laboratory | RTL |
| refining | refg |
| reflect (ed) (ing) (tion) | refl, RFLCT |
| reflux | refl |
| reformate (er) (ing) | reform |
| reformer | REFMR |
| refraction, refractory | refr |
| refrigeration building | RFG/BD |
| refrigerator (rant) (tion) | REFRIG |
| regenerator | regen |
| register | reg, RGTR |
| regular acid | R/A |
| regular, regulator | reg |
| Reid vapor pressure | RVP |
| reinforce (ed) (ing) (ment) | reinf |
| reinforced concrete | reinf conc |
| reinforcing bar | rebar |
| reject | rej |
| rejection | rej'n |
| Reklaw | Rek |
| relative humidity | RH |

| | |
|---|---|
| relay | rly |
| release (ed) (ing) | rls (ed) (ing), red |
| released swab unit | RSU |
| relief | rlf |
| relief valve | RV |
| relocate (ed) | reloc |
| remains | rmns, rems |
| remedial | rem |
| remote control | RC |
| remote operating system (station) | ROS |
| remote terminal unit | RTU |
| remove (al) (able) | rmv (l) |
| Renault | Ren |
| rental | rent |
| Reophax bathysiphoni | Reo. bath. |
| repair (ed) (ing) (s) | rep |
| repairman | rpmn |
| reperforated | reperf |
| replace (ed) | rep |
| replace (ment) | repl |
| report | rep, RPRT |
| reported | rptd |
| request | REQ |
| request for proposal | RFP |
| request for quote | RFQ |
| required | reqd |
| requirement | reqmt |
| requisition | req |
| research | res |
| research and development | R&D |
| Research Octane Number | Res. O. N., RON |
| Research Planning Institute | RPI |
| reserve (ation) | res |
| reservoir | rsvr |

## Definitions with Abbreviations

| | |
|---|---|
| reservoir description service | RDS |
| residual, residue | resid |
| resinous | rsns |
| resistance, resistivity, resistor | res |
| resistivity | R |
| resistivity (as recorded from 16" electrode configuration) | R(16") |
| resistivity invaded zone | RIZ |
| resistivity, flushed zone | Rxo |
| resistivity, mud | Rm |
| resistivity, mud filtrate | Rmf |
| resistivity, water | Rw |
| resistivity, water (apparent) | Rwa |
| resistor (s) | RESIS |
| retail pump price | RPP |
| retain (er) (ed) (ing) | ret, rtnr |
| retard (ed) | rtd |
| retrievable | retr |
| retrievable bridge plug | RBP |
| retrievable retainer | retr ret |
| retrievable test treat squeeze (tool) | RTTS |
| return | ret |
| return on investment | ROI |
| returnable steel drum | RSD |
| returned | retd |
| returned well to production | RWTP |
| returning circulation oil | RCO |
| reverse (ed) | res, rev (d) |
| reverse circulation | RC |
| reverse circulation rig | RCR |
| reversed out | rev/O, RO |
| revise (ed) (ing) (ion) | rev |
| revolution (s) | rev |
| revolutions per minute | rpm |

| | |
|---|---|
| revolutions per second | rps |
| rework (ed) | rwk (d) |
| rheostat | rheo |
| ribbon sand | Rib |
| rich oil fractionator | ROF |
| Rierdon rig | Rier |
| rig (ged) (ging) up | RU |
| rig floor | RF |
| rig on location | ROL |
| rig released | RR, R Rel |
| rig repair | RR |
| rig service | RS |
| rig skidded | RS |
| rig time | RT |
| rig-up casing crew | RUCC |
| rigged (ing) down | RD |
| rigged down, moved out | RDMO |
| rigged-down swabbing unit | RDSU |
| rigging rotary | RR |
| rigging-up cable tools | RUCT |
| rigging-up machines | RUM |
| rigging-up pump | RUP |
| rigging-up rotary tools | RURT |
| rigging-up service rig | RUSR |
| rigging-up standard tools | RUST |
| rigging-up swabbing unit | RUSU |
| rigging-up tools | RUT |
| right angle | RA |
| right hand | RH |
| righthand door | RHD |
| right of way | ROW |
| ring | rg |
| ring groove | RG |
| ring joint | RJ |

| | |
|---|---|
| ring joint, flanged end | RJFE |
| ring tool joint | RTJ |
| ring-type joint | RTJ |
| rising stem (valve) | RS |
| rivet | riv., RVT |
| road (s) | rd (s) |
| road & location | R&L |
| road & location complete | R&LC |
| Robulus | Rob |
| rock | rk |
| rock bit | RB |
| rock pressure | RP |
| Rockwell hardness number | RHN |
| rocky | rky |
| Rodessa | Rod |
| rods and tubing | R&T |
| roentgen | r |
| roofing | RFG |
| room | rm |
| root mean square | RMS |
| rose | ro |
| Rosiclare sand | Ro sd |
| rotameter | RTMTR |
| rotary bushing | RB |
| rotary bushing measurement | RBM |
| rotary drive bushing | RDB |
| rotary drive bushing to ground | RDB-GD |
| rotary kelly bushing | RKB |
| rotary time | RT |
| rotary test | R test |
| rotary tools | RT |
| rotary total depth | RTD |
| rotary unit | RU |
| rotary, rotate, rotator | rot |

| | |
|---|---|
| rotative gas lift | ROGL |
| rough | rgh |
| rough order of magnitude | ROM |
| round | rd |
| round thread | rd thd |
| round trip | rdtp |
| round trip changed bit | RT CB |
| rounded | rdd, rnd |
| routing | RTG |
| rows | R |
| royalty | roy |
| royalty interest | RI |
| rubber | rbr, rub |
| rubber ball sand water frac | RBSWF |
| rubber ball sand oil frac | RBSOF |
| rubber balls | Rbls |
| run of mine | ROM |
| running | rng |
| running casing | RC |
| running electric log | REL |
| running radioactive log | RALOG |
| running tubing | RTG |
| rupture | rupt |
| rust and oxidation | R&O |

## S

| | |
|---|---|
| Sabinetown | Sab |
| saccharoidal | sach |
| sack (s) | sk, sx |
| saddle | sadl |
| Saddle Creek | Sad Cr |
| safety | saf |

## Definitions with Abbreviations

| | |
|---|---|
| safety relief valve | SRV |
| safety/department | SAF/DPT |
| Saint Genevieve | St Gen |
| Saint Louis lime | St L |
| Saint Peter | St Ptr |
| Salado | Sal |
| salary, salaried | sal |
| Saline Bayou | Sal Bay |
| salinity | sal |
| salt | X |
| salt and pepper | s&p |
| Salt Mountain | Slt Mtn |
| salt wash | SW |
| salt water | SW, swtr, XW |
| salt water to surface | SWTS |
| saltwater-cut mud | SWCM |
| saltwater disposal | SWD |
| saltwater disposal system | SWDS |
| saltwater disposal well | SWDW |
| saltwater fracture | SWF |
| saltwater injection | SWI |
| salty | Slty |
| salty sulfur water | SSUW |
| salvage | salv |
| sample | samp, smpl, spl |
| sample chamber | splcham |
| sample formation tester | SFT |
| sample tops | S/T |
| San Andres | San And |
| San Angelo | San Ang |
| San Bernardino base and meridian | SBB&M |
| San Rafael | San Raf |
| Sanstee | Sana |
| sand | SD, sd |

| | |
|---|---|
| sand and shale | sd&sh |
| sand-oil fracked | sdoilfract |
| sand-oil fracture | SOF |
| sand showing gas | Sd SG |
| sand showing oil | Sd SO |
| sand-water fracked | sdwtrfract |
| sanded | sdd |
| sandfracked | sdfract, SF |
| sandstone | SS |
| sandy | sdy |
| sandy lime | sdy li |
| sandy shale | sdy sh |
| sanitary | SAN, sani |
| sanitary water | S/WTR |
| Santa Margarita | Sta Marg |
| saponification | sap |
| saponification number | Sap No. |
| Saratoga | Sara |
| Satanka | Stnka |
| saturated, saturation | sat |
| Sawatch | Saw |
| Sawtooth | Sawth |
| Saybolt furol | Say Furol |
| Saybolt Seconds Universal | SSU |
| Saybolt universal viscosity | SUV |
| scaffolding | SCAF |
| scales | sc |
| scatter (ed) | scat (d), sctr (d) |
| schedule | sch |
| schematic | schem |
| Schlumberger | SCHL |
| scolescodonts | scolc |
| scraper | scr |
| scratcher | scr |

## Definitions with Abbreviations

| | |
|---|---|
| screen | scr |
| screw (ed) | scr (d) |
| screw end American National Acme thread | SE NA |
| screw end American National Coarse thread | SE NC |
| screw end American National Taper Pipe thread | SE NTP |
| screwed and socketweld | S/SW |
| screwed end | S/E |
| screwed on one end | SOE |
| scrubber | scrub |
| sea level | S.L. |
| Seabreeze | Sea |
| seal assembly | SA |
| seal oil | SEO |
| seal weld | SWLD |
| seal-welded bonnet | SWB |
| sealed | sld |
| seamless | smls |
| seating nipple | SN |
| secant | sec |
| second (ary) | sec |
| secondary butyl alcohol | SBA |
| seconds | S, sec |
| secretary | sec |
| section | sec |
| section (s) (al) (ing) | SECT |
| section line | SL |
| section-township-range | S-T-R |
| securaloy | scly |
| sediment (s) | sed |
| Sedwick | Sedw |
| segment | SEG |
| seismograph, seismic | seis |
| selection (tive) (tor) | SELECT |
| selenite | sel |

259

| | |
|---|---|
| self (spontaneous) potential | SP |
| self-contained | SC |
| self-elevating work platform | SEWOP |
| Selma | Sel |
| Senora | Sen |
| separate, separator, separation | SEP |
| septuplicate | sept |
| sequence | seq |
| series, serial | ser |
| serpentine | serp |
| Serratt | Serr |
| service | svc, serv |
| service charge | serv chg |
| service unit | svcu |
| set drillpipe | SDP |
| set plug | SP |
| settling | set |
| Seven Rivers | S Riv |
| service station | SS |
| severy | Svry |
| Seward Meridian (Alaska) | SM |
| sewer | sew |
| sexton | Sex |
| sextuple | sxtu |
| sextuplicate, sextuplet | sext |
| shaft horsepower | shp |
| shake out | SO |
| shale | sh |
| shaled out | SO |
| shaly | shly |
| shallower pool (pay) test | SPT |
| shallower pool wildcat, discovery | WSD |
| shallower pool wildcat, dry | WS |
| Shannon | Shan |

## Definitions with Abbreviations

| | |
|---|---|
| shear | shr |
| sheathing | shthg |
| sheet | sh |
| sheeting | SHTG |
| shell and tube | S&T |
| shells | shls |
| shelter | SHLT |
| Shinarump | Shin |
| ship (ping) | shp(g) |
| shipping point (purchasing term) | s/p, sp |
| shipment | shpt |
| shock sub | SS |
| shop fabrication | S/FAB |
| short radius | SR |
| short string | SS |
| short thread | ST |
| short threads & coupling | ST&C |
| shortage | SHTG |
| shot open hole | SOH |
| shot per foot | SPF |
| shot point | sp |
| shoulder | shld |
| show condensate | SC |
| show gas | SG |
| show gas and condensate | SG&C |
| show gas and distillate | SG&D |
| show gas and water | SG&W |
| show of dead oil | SDO |
| show of free oil | SFO |
| show gas and oil | SG&O |
| show of oil side opening | SO |
| show oil and gas | SO&G |
| show oil and water | SO&W |
| shut down | SD |

261

| | |
|---|---|
| shut down awaiting orders | SDWO |
| shut down for orders | SDO |
| shut down for night | SDFN |
| shut down for pipeline | SDPL |
| shut down for repairs | SDR |
| shut down for weather | SDW |
| shut down overnight | SDON |
| shut down to acidize | SDA |
| shut down to fracture | SDF |
| shut down to log | SDL |
| shut down to plug and abandon | SDPA |
| shut in | SI |
| shutin bottom-hole pressure | SIBHP |
| shutin casing pressure | SICP |
| shutin gas well | SIGW |
| shutin oil well | SIOW |
| shutin pressure | SIP |
| shutin tubing pressure | SITP |
| shutin wellhead pressure | SIWHP |
| shutin, waiting on potential | SIWOP |
| shut well in overnight | SWION |
| side door choke | SD Ck |
| side opening | SO |
| sidewall cores | SWC |
| sidewall neutron porosity | SWNP |
| sideboom | SB |
| siderite (ic) | sid |
| sides, tops, & bottoms | s, t&b |
| sidetrack (ed) (ing) | sdtkr, ST |
| sidetracked hole | STH |
| sidetracked total depth | STTD |
| sidewall | SDWL |
| sidewall samples | SWS |
| siding | SDG |

## Definitions with Abbreviations

| | |
|---|---|
| signed | sgd |
| silencer | SLNCR |
| silica, siliceous | silic |
| silky | slky |
| siltstone | silt |
| silty | slty |
| Silurian | Sil |
| similar | sim |
| Simpson | Simp |
| single (s) | sgl (s) |
| single-pole double throw | SPDT |
| single-pole double throw switch | SPDT SW |
| single-pole single throw | SPST |
| single-pole single throw stitch | SPST SW |
| single-pole switch | SP SW |
| single random lengths | SRL |
| single shot | SS |
| Siphonina davisi | Siph. d. |
| size | sz |
| sketch | SK |
| skimmer | skim |
| Skinner | Skn |
| Skull Creek | Sk Crk |
| sleeper | SLPR |
| sleeve | sl, SLV |
| sleeve bearing | SB |
| slickensided | sks |
| sliding-scale royalty | S/SR |
| slight (ly) | sli |
| slight oil-cut mud | SOCM |
| slight oil-cut water | SOCW |
| slight show of gas | SSG |
| slight show of oil | sli SO, SSO |
| slight show of oil and gas | SSO&G |

## Standard Oil & Gas Abbreviator

| | |
|---|---|
| slight, weak, or poor fluorescence | SFLU |
| slightly gas-cut salt water | SGCSW |
| slightly gas-cut water cushion | SGCWC |
| slightly gas-cut mud | SGCM |
| slightly gas-cut oil | SGCO |
| slightly gas-cut water | SGCW |
| slightly gas-cut water blanket | SGCWB |
| slightly oil- and gas-cut mud | SO&GCM |
| slight oil-cut salt water | SOCSW |
| slight oil-cut water blanket | SOCWB |
| slight oil-cut water cushion | SOCWC |
| slightly porous | sp |
| Sligo | Sli |
| slim-hole drillpipe | SHDP |
| slip and cut drill time | SC DL |
| slip on | SO |
| slope type of wall to keep out flooding | berm |
| slow set (cement) | SS |
| slurry | slur |
| Smackover | Smk, SO |
| small | sm |
| small show | SS |
| Smithwick | Smithw |
| smoke volatility index | SVI |
| smooth | smth |
| snubber, snubbing | SNUB |
| snuffing | SNUFF |
| Society of Economic Paleontologists & Mineralogists | SEPM |
| Society of Petroleum Engineers | SPE |
| socket | skt |
| socket weld | SW |
| sodium-based grease | sod gr |
| sodium carbonate | $NaCO_3$ |

| | |
|---|---|
| sodium carboxymethylcellulose | CMC |
| sodium chloride | NaCL |
| sodium hydroxide | NaOH |
| soft | sft |
| solar heat medium | SHM |
| solenoid | sol, slnd |
| solenoid-operated valves | SOV |
| solenoid valve | SV |
| solids | sol |
| solution | soln |
| solvent | solv |
| somastic | som |
| somastic coated | somct |
| sonic log | SONL |
| sort (ed) (ing) | srt |
| south | S |
| south half | S/2 |
| south line | SL |
| south offset | S O |
| southeast | SE |
| southeast corner | SE/C |
| southeast quarter | SE/4 |
| southerly | S'ly |
| southwest corner | SW/c |
| southwest quarter | SW/4 |
| spacer | spcr |
| spare | sp |
| Sparta | Sp |
| spearfish | spf |
| special | spcl |
| specialty | splty |
| specific gravity | sp gr |
| specific heat | sp ht |
| specific volume | sp. vol. |

| | |
|---|---|
| specification | spec |
| speckled | speck |
| speed/current | S/C |
| speed/torque | S/T |
| Sphaerodina | Sphaer |
| sphalerite | sphal |
| spherules | sph |
| spicule (ar) | spic |
| spigot | SPGT |
| spigot and spigot | s&s |
| spillway | SPWY |
| spindle | spdl |
| Spindletop | spletp |
| spiral weld | SW |
| spirifers | sprf |
| Spiroplectammina barrowi | Spiro. b. |
| splintery | sply |
| splitter | SPLTR |
| sponge | spg |
| spore | sp |
| spot sales agreement | SSA |
| spotted | sptd |
| spotty | sptty |
| Spraberry | Spra |
| spring | spg |
| Springer | Sprin |
| sprinkler | spkr |
| sprocket | spkt |
| spud (ded) (der) | spd |
| square | sq |
| square centimeter | sq cm |
| square foot | ft$^2$, sq ft |
| square inch | sq in. |
| square kilometer | sq km |

## Definitions with Abbreviations

| | |
|---|---|
| square meter | sq m |
| square millimeter | sq mm |
| square root | SQRT |
| square yard (s) | sq yd |
| squeeze (ed) (ing) | sqz |
| squeeze packer | sq pkr |
| squeezed | sq |
| squirrel cage | sq cg |
| stabilized (er) | stab |
| stage | STG |
| staggered | STAG |
| stain (ed) (ing) | stn (d) (g) |
| stain and odor | S&O |
| stainless steel | SS |
| stairway | stwy |
| Stalnacker | Stal |
| stand (s) (ing) | std |
| stand by | stn/by |
| standard cubic feet per day | SCFD, scfd, |
| standard cubic feet per hour | SCFH, scfh |
| standard cubic feet per minute | SCFM, scfm |
| standard cubic foot | SCF, scf |
| standard operational procedure | SOP |
| standard temperature and pressure | STP |
| standardization | STDZN |
| standards | std (s) |
| standing | stdg |
| Stanley | Stan |
| start | st |
| start of cycle | SOC |
| start of run | SOR |
| started in hole | SIH |
| started out of hole | SOH, SOOH |
| starting fluid level | SFL |

267

| | |
|---|---|
| state lease | SL |
| state potential | State pot |
| statewide rules | SWR |
| static bottom-hole pressure | SBHP |
| station | sta |
| stationary | stat |
| statistical | stat |
| steady | stdy |
| steam | stm |
| steam cylinder oil | stm cyl oil |
| steam emulsion number | SE No. |
| steam engine oil | stm eng oil |
| steam trace (ing) | STM TR |
| steam working pressure | SWP |
| steel | stl |
| steel line correction | SLC |
| steel line measurement | SLM |
| steel tape measurement | STM |
| Steele | Stel |
| stencil (ed) (ing) | stncl (d) (g) |
| stenographer | steno |
| Stensvad | Stens |
| sticky | stcky |
| stiffener | STIF |
| stippled | stip |
| stirrup | stir |
| stock | stk |
| stock tank barrels | STB, stb |
| stock tank barrels per day | STBPD, stb/d |
| stock tank oil in place | st oip |
| stock tank vapor | STV |
| Stone Corral | Stn Crl |
| Stony Mountain | Sty Mtn |
| stopper (rd) | stpr (d) |

| | |
|---|---|
| storage | strg |
| stove oil | stv |
| straddle | strd |
| straddle packer | SP |
| straddle packer drillstem test | SP-DST |
| straight | strt |
| straight-hole test | SHT |
| straight-run naphtha | SRN |
| straightened | strtd |
| straightening | stging |
| strainer | stnr |
| strand (ed) | strd |
| strap out of hole | STROH |
| strapped out of hole | SOOH |
| stratigraphic | strat |
| Strawn | Str |
| streak | stk, strk |
| striated | stri |
| string (er) | strg (r) |
| string shot | SS |
| strip (per) (ping) | STPR |
| strokes per minute | SPM |
| stromatoporoid | strom |
| strong | strg |
| structure, structural | struc |
| stuck | stl |
| study | stdy |
| stuffing box | SB |
| styolite, styolitic | styo |
| sub-Clarksville | Sub Clarks |
| subangular | sub angl |
| subdivision | subd |
| subrounded | sub rnd |
| subsea | SS |

269

| | |
|---|---|
| subsidiary | sub |
| substance | sub |
| substation | substa |
| substitute | SUBST |
| substructure height | SH |
| subsurface | SS |
| subsurface safety valve | SSSV |
| successful wildcat outpost | WOE |
| sucker rod | skr d |
| sucrose, sucrosic | suc |
| suction | suct |
| sugary | sug |
| sulfate bacteria | SRB |
| sulfur by bomb method | S Bomb |
| sulfur, sulfuric | sulf |
| sulfuric acid | $H_2SO_4$ |
| sulfated | sulf |
| sulfur | sul |
| sulfur water | sul wtr |
| summary, summarize | sum |
| Summerville | Sumvl |
| Sunburst | Sb, Sunb |
| Sundance | Sund |
| Supai | Sup |
| superintendent | supt |
| superseded | supsd |
| supervisor | suprv |
| supplement | supp |
| supply (ied) (ier) (ing) | supl, sply |
| support | suppt |
| surface | surf, sfc |
| surface approximation and formation evaluation | SAFE |
| surface-controlled subsurface safety valve | SCSSV |

| | |
|---|---|
| surface flow pressure | SFP |
| surface geology | SG |
| surface measurement | SM |
| surface pressure | SP |
| surge | SRG |
| surplus | sur |
| survey | sur |
| suspended | susp |
| suspended ceiling | SUSP CLG |
| swab and flow | S&F |
| swab rate | SR |
| swab run (s) | SR |
| swabbed | S/ |
| swabbed, swabbing | SWB |
| swabbing unit | SWU |
| swaged | swd |
| Swastika | Swas |
| sweetening | Swet |
| switch | SW |
| switchboard | PBX, swbd |
| switchgear | swgr |
| switchrack | SWRK |
| Sycamore | Syc |
| Sylvan | Syl |
| symbol | sym |
| symmetrical | sym |
| synchronous, synchronizing | syn |
| synchronous converter | syn conv |
| synchroscope | SYNSCP |
| synthesis | SYNTH |
| synthetic | syn |
| synthetic natural gas | SNG |
| system | sys |
| system flow diagram | SFD |

# T

| | |
|---|---|
| tabular, tabulating | tab |
| tachometer | TACH |
| tag closed cup (flash) | TCC |
| tag open cup (flash) | TOC |
| Tagliabue | Tag |
| Tallahatta | Tal |
| Tampico | Tamp |
| tangent | tan |
| tank | tk |
| tank battery | TKB |
| tank car | T/C |
| tank farm | TKF |
| tank truck | TT |
| tank wagon | TKW |
| tankage | tkg |
| tanker (s) | tkr |
| Tannehill | Tann |
| Tansill | Tan |
| taper pipe thread | TPT |
| Tar Springs sand | TSS |
| Tarkio | Tark |
| tarred and wrapped | T&W |
| taste | tste |
| Taylor | Tay |
| tearing out rotary tools | TORT |
| technical, technician | tech |
| techniques of water-resources investigations | TWI |
| tee | T |
| teeth | T |
| telegraph | TLG |
| Telegraph Creek | Tel Cr |
| teletype | TWX |

## Definitions with Abbreviations

| | |
|---|---|
| television | TV |
| telephone, telegraph | tel |
| temperature | Temp |
| temperature control valve | TCV |
| temperature controller | TC |
| temperature differential indicator | TDI |
| temperature differential recorder | TDR |
| temperature gradient | TG |
| temperature indicator | TI |
| temperature indicator controller | TIC |
| temperature observation | TO |
| temperature recorder | TR |
| temperature recorder controller | TRC |
| temperature survey indicated top cement at | TSITC |
| temporarily abandoned | TA |
| temporarily shutdown | TSD |
| temporarily shut in | TSI |
| temporary (ily) | temp |
| temporary dealer allowance | TDA |
| temporary voluntary allowance | TVA |
| tender | tndr |
| tensile strength | tens str, TS |
| tension leg platform | TLP |
| Tensleep | Tens |
| tentaculites | Tent |
| tentative | tent |
| Teremplealeau | Tremp |
| terminal board | T/BRD |
| terminal box | T/Box |
| terminate (ed) (ing) (ion) | termin |
| tertiary | Ter |
| tertiary butyl alcohol | TBA |
| test (er) (ing) | tst (r) (g) |
| test to follow | TTF |

273

Standard Oil & Gas Abbreviator

| | |
|---|---|
| testing on pump | TOP |
| tetraethyl lead | TEL |
| tetramethyl lead | TML |
| Texana | Tex |
| Textularia articulate | Text. art. |
| Textularia dibollensis | Text. d. |
| Textularia hockleyensis | Text. h. |
| Textularia warreni | Text. w. |
| texture | tex |
| Thaynes | Thay |
| thence | th |
| theoretical production and allocation | TP&A |
| thermal | thrm |
| thermal crack | therm ckr |
| thermal decay time | TDT |
| thermal hydrodealkylation | THD |
| thermofor catalytic cracking | TCC |
| thermometer | therm |
| Thermopolis | Ther |
| thermostat | therst |
| thick, thickness | thk |
| thin bedded | TB |
| thousand (i.e., 133K = 13,000) | K |
| thousand (i.e., 9M = 9,000) | M |
| thousand barrels fluid per day | MBF/D, MBFPD |
| thousand barrels of oil per day | MBO/D, MBOPD |
| thousand barrels of water per day | MPW/D, MBWPD |
| thousand British thermal units | MBtu |
| thousand cubic feet | MCF |
| thousand cubic feet of gas per day | MCFGPD, Mcfgpd |
| thousand cubic feet per day | MCFD, Mcfd |
| thousand electron-volts | kev |
| thousand gallons | MG |
| thousand standard cubic feet | MCSF, Mscf |

| | |
|---|---|
| thousand standard cubic feet per day | MSCF/D, Mscf/d |
| thousand standard cubic feet per hour | MSCF/H, Mscf/h |
| thread large end | TLE |
| thread on both ends | TOBE |
| thread small end | TSE |
| thread small end, weld large end | TSE-WLE |
| thread, threaded | thd |
| threaded and coupled | T&C |
| threaded both ends | TBE |
| threaded one end | TOE |
| threaded pipe flange | TPF |
| Three Finger | Tfing |
| Three Forks | Tfks |
| three-phase | 3 PH |
| throttling | thrling |
| through | thru |
| through the tanks | TTT |
| through-tubing | TT |
| through-tubing caliber | TTC |
| through-tubing plug | TTP |
| Thurman | Thur |
| tight | ti, tite |
| tight hole | TH |
| tight no show | TNS |
| time delay | TD |
| Timpas | TIM |
| Timpoweap | Timpo |
| tires, batteries, and accessories | TBA |
| to be conditioned for gas | TRG |
| to be conditioned for oil | TRO |
| Todilto | Tod |
| tolerance | tol |
| toluene | tolu |
| ton (after number-3T) | T |

| | |
|---|---|
| tongue and groove (joint) | T&G |
| Tonkawa | Tonk |
| tons | tons |
| too small to measure | TSTM |
| too wet (weak) to measure | TWTM |
| tool (s) | tl |
| tool closed | TC |
| tool joint | tl jt |
| tool open | TO |
| tool pusher | TP |
| tooth, teeth | T |
| top and bottom | T&B |
| top and bottom chokes | T&BC |
| top choke | TC |
| top hole choke | THC |
| top hole flow pressure | THFP |
| top of (a formation) | T/ |
| top of cement | TOC |
| top of cement plug | TOCP |
| top of fish | TOF |
| top of liner | TOL |
| top of liner hanger | TLH |
| top of pay | T/pay |
| top of salt | TOS |
| top of sand | T/sd |
| top salt | T/S |
| Topeka | Tpka |
| topo sheet evaluation | TS |
| topographic, topography | topo |
| topping | topg |
| topping and coking | T&C |
| Toronto | Tor |
| Toroweap | Toro |
| torque | TRQ |

## Definitions with Abbreviations

| | |
|---|---|
| total | tot |
| total depth | TD |
| total time lost | TTL |
| totally enclosed, fan cooled | TEFC |
| tough | gh |
| Towanda | Tow |
| tower | TWR |
| township | twp |
| township (as T2N) | T |
| townsite | twst |
| trace | TR |
| trackage | trkg |
| tract | TR |
| Trans-Alaska Pipeline System | TAPS |
| transducer | TRNDC |
| transfer (ed) (ing) | trans |
| transformer | trans, |
| translucent | transl |
| transmission | trans |
| transmitter | XTMR |
| transparent | transp |
| transportation | transp |
| travel (ed) (ing) | TRVL |
| Travis Peak | TP |
| treat (er) (ed) (ing) | trt (r)(d)(g) |
| treater | trtr |
| treating pressure | TP |
| Trenton | Tren |
| Triassic | Tri |
| tricresyl phosphate | TCP |
| trillion | $10^{12}$ |
| trillion cubic feet | Tcf, TCF |
| trillion cubic feet per day | TCF/D, Tcf/d |
| trilobite | trilo |

277

| | |
|---|---|
| Trinidad | Trin |
| trip for new bit | TFNB |
| trip in hole | TIH |
| trip out of hole | TOH |
| trip (ped) for bit | TFB |
| triple-pole single throw switch | 3P ST SW |
| triple-pole switch | 3P SW |
| triplicate | trip |
| Tripoli | Trip |
| tripolitic | trip |
| tripped (ing) | trip |
| truck | trk |
| true boiling point | TBP |
| true vapor process | TVP |
| true vertical depth | TVD |
| tube | tb |
| tube bundle | TB/BDL |
| tubing | tbg |
| tubing and casing cutter | TCC |
| tubing and rods | T&R |
| tubing choke | tbg chk, TC |
| tubing pressure | TP, tbg press |
| tubing pressure shut in | TPSI |
| tubing pressure, closed | TPC |
| tubing pressure, flowing | TPF |
| tubinghead flange | THF |
| Tucker | Tuck |
| tuffaceous | tfs, tuf |
| Tulip Creek | Tul Cr |
| tungsten carbide | tung carb |
| turbine compressor | T/C |
| turbo, turbine | TURB |
| turn around | TA |
| turned over to producing section | TOPS |

Definitions with Abbreviations

| | |
|---|---|
| turned to test tank | TTTT |
| turnpike | tpk |
| Tuscaloosa | Tus |
| 12 gauge wire-wrapped screen (in a liner) | 12GA W.W.S. |
| Twin Creek | Tw Cr |
| twisted off | twst off |
| type | ty |
| typewriter | tywr |
| typical | typ |

| | |
|---|---|
| ultimate | ult |
| ultra-high frequency | UHF |
| ultrasonic examination | UT |
| ultrasonic test | UST |
| ultraviolet | UV |
| umbrella (s) | UMB |
| Umiat Meridian (Alaska) | UM |
| unbalanced | UNBAL |
| unbranded | unbr |
| unclassified | U |
| unconformity | unconf |
| unconsolidated | uncons |
| under construction | U/C |
| under digging | UD |
| under gauge | UG |
| underreaming | UR |
| underground | UG |
| undifferential | undiff |
| unfinished | unf |
| unified coarse thread | UNC |
| unified fine thread | UNF |

279

| | |
|---|---|
| uniform | uni |
| uninterruptible power supply | UPS |
| Union Oil Company | UOCO |
| Union Valley | UV |
| unit | un |
| United States gauge | USG |
| universal gear lubricant | UGL |
| universal transverse mercator | UTM |
| university, universal | Univ |
| unloader | UNLDR |
| unloading | UNLD |
| unsulfonated residue | UR |
| upper (i.e., U/Simpson) | U/ |
| upper and lower | U/L |
| upper casing | UC |
| upper tubing | UT |
| upthrown | UT |
| use customer's hose | UCH |
| used rod | UR |
| used with | U/W |
| utility | UTL |
| utility flow diagram | UFD |
| utility water | U/WTR |
| Uvigerina lirettensis | Uvig. lir. |

## V

| | |
|---|---|
| vacant | vac |
| vacation | vac |
| vacuum | vac |
| Vaginuline regina | Vag. reg. |
| Valera | Val |
| valve | V, vlv |
| Vanguard | Vang |

## Definitions with Abbreviations

| | |
|---|---|
| vapor pressure | VP |
| vapor recovery | VR |
| vapor recovery unit | VRU |
| vapor temperature | vt |
| vapor (izor) | vap (r) |
| vapor-liquid ratio | V/L |
| varas | vrs |
| variable, various | var |
| variegated | vari |
| varnish | VARN |
| varnish makers and painters naphtha | VM&P naptha |
| varved | vrvd |
| velocity | vel |
| velocity survey | V/S, VS |
| vendor drawing | V/DWG |
| ventilator | vent |
| Verdigris | Verd |
| Vermillion Cliff | Ver Cl |
| versus | vs |
| vertebrate | vrtb |
| vertical | vert, vrtl |
| vertical support member | VSM |
| very (as very tight) | v. |
| very common | v.c. |
| very fine-grain (ed) | vfg |
| very heavily (highly) gas-cut mud | VHGCM |
| very heavily (highly) gas-cut salt water | VHGCSW |
| very heavily (highly) gas-cut water | VHGCW |
| very heavily (highly) oil- and gas-cut salt water | VHO&GCSW |
| very heavily (highly) oil- and gas-cut mud | VHO&GCM |
| very heavily (highly) oil- and gas-cut water | VHO&GCW |
| very heavily (highly) oil-cut mud | VHOCM |
| very heavily (highly) oil-cut salt water | VHOCSW |
| very heavily (highly) oil-cut water | VHOCW |

| | |
|---|---|
| very high frequency | VHF |
| very light amber cut | VLAC |
| very noticeable | v.n. |
| very poor sample | V.P.S. |
| very rare | v.r. |
| very slight | v-sli |
| very slight show of gas | VSSG |
| very slight show of oil | VSSO |
| very slightly gas-cut salt water | VSGCSW |
| very slightly gas-cut water | VSGCW |
| very slightly gas-cut mud | VSGCM |
| very slightly oil- and gas-cut mud | VSO&GCM |
| very slightly oil- and gas-cut salt water | VSO&GCSW |
| very slightly oil-cut mud | VSOCM |
| very slightly oil-cut salt water | VSOCSW |
| very slightly oil-cut water | VSOCW |
| very slightly porous | VSP |
| vesicular | ves |
| vessel | VESS |
| vibrate (tor) (ing) | VIB |
| Vicksburg | Vks |
| Viola | Vi |
| Virgelle | Virg |
| viscosity | vis, V |
| viscosity index | VI |
| viscosity-gravity constant | VGC |
| visible | vis |
| vitreous | vit |
| vitrified clay pipe | VCP |
| Vogtsberger | Vogts |
| volatile organic compounds | VOC |
| volt | V, v |
| Volt-ampere | va |
| volt-ampere reactive | var |

Definitions with Abbreviations

| | |
|---|---|
| voltage | VOLT |
| volume | V, vol |
| volume-percent | v% |
| volumetric efficiency | vol. eff. |
| volumetric subcommittee | VSC |
| vuggy | vug |
| vugular | vug |

| | |
|---|---|
| Wabaunsee | Wab |
| Waddell | Wad |
| waiting | wtg |
| waiting on | WO |
| waiting on acid | WOA |
| waiting on allowable | WOA |
| waiting on battery | WOB |
| waiting on cable tools or completion tools | WOCT |
| waiting on completion rig | WOCR |
| waiting on drillpipe | WODP |
| waiting on geologist | WOG |
| waiting on orders | WOO |
| waiting on permit | WOP |
| waiting on pipe | WOP |
| waiting on pipeline | WOPL |
| waiting on plastic | WOP |
| waiting on potential test | WOPT |
| waiting on production equipment | WOPE |
| waiting on pump | WOP |
| waiting on pumping unit | WOPU |
| waiting on rig or rotary | WOR |
| waiting on rotary tools | WORT |
| waiting on standard tools | WOST |
| waiting on state potential | WOSP |

283

| | |
|---|---|
| waiting on tank and connection | WOT&C |
| waiting on test or tools | WOT |
| waiting on weather | WOW |
| wall (if used with pipe) | W |
| Wall Creek | W Cr |
| wall thickness (pipe) | WT |
| Waltersburg sand | Wa sd |
| Wapanucka | Wap |
| warehouse | whse |
| Warsaw | War |
| Wasatch | Was |
| wash (ing) | wsh (g) |
| wash and ream | W&R |
| wash and ream to bottom | WRTBw |
| wash oil | WO |
| wash over | WO |
| wash pipe | WP |
| wash to bottom | WTB |
| wash water | WW |
| washed | w shd |
| washing in | Wl |
| Washita | Wash |
| Washita-Fredericksburg | W-F |
| washout | wo |
| washover string | WOS |
| waste | WSTw |
| water blanket | WB |
| water closet | WC |
| water cooler | W/CLR |
| water cushion (DST) | wtr. cush, WC |
| water cushion to surface | WCTS |
| water cut | WC |
| water depth | WD |
| water disposal well | WD |

## Definitions with Abbreviations

| | |
|---|---|
| water in hole | WIH |
| water injection | Wl |
| water injection well | WIW |
| water load | W/L |
| water loss | WL |
| water not shut off | WNSO |
| water oil or gas | WOG |
| water saturation | WS |
| water separation index modified | WSIM |
| water shutoff no good | WSONG |
| water shutoff OK | WSOOK |
| water shutoff | WSO |
| water source wells | WST |
| water supply well | WSW |
| water to surface | WTS |
| water well | WW |
| water with slight show of oil | W/SSO |
| water with sulfur odor | W/sulf O |
| water-alternating gas (or water and gas) | WAG |
| water-cut mud | WCM |
| water-cut oil | WCO |
| water-oil ratio | WOR |
| water-supply paper | WSP |
| water-up to | WUT |
| water, watery | wtr (y) |
| waterflood | WF |
| waterproof | WTR/PRF |
| watertight | WTR/T |
| wating on cement | WOC |
| watt | w |
| watt-hour | w-hr |
| weak | wk |
| weak air blow | WAB |
| weather | wthr |

| | |
|---|---|
| weather (ed) | wthrd |
| weatherproof | WTH/PRF |
| Weber | Web |
| week | wk |
| weight | wgt., wt |
| weight averaged catalyst temperature | WACT |
| weight on bit | W.O.B. |
| weight-percent | w% |
| Weighted Average Cost of Gas | WACOG |
| welded steel tank | W/S TK |
| weld ends | WE |
| weldneck | WN |
| welded, welding | wld |
| welder | wldr |
| welding detail (s) | WLD/DET |
| Welex | Wx |
| wellhead injection pressure | WHIP |
| well lines | WL |
| well pad | WP |
| well pad manifolding | WPM |
| wellbore | wlbr |
| wellhead | WH |
| Wellington | Well |
| went back in hole | WBIH |
| west | W |
| west half | W/2 |
| west line | WL |
| west offset | W/O |
| westerly | W'ly |
| wet bulb | WB |
| wheel | WHL |
| whipstick | whip, WS |
| whipstock depth | WSD |
| white | wht |

| | |
|---|---|
| white dolomite | Wh Dol |
| White River | WR |
| white sand | Wh Sd |
| wholesale | whsle |
| Wichita | Wich. |
| Wichita Albany | Wich Alb |
| wide | W |
| wide flange | WF |
| Wilcox | Wx |
| wildcat | WC |
| wildcat field, discovery | WFD |
| wildcat outpost, dry | WO |
| Willberne | Willb |
| Wind River | Wd R |
| Windfall Profit Tax | WPT |
| Winfield | Winf |
| Wingate | Wing |
| Winnipeg | Winn |
| Winona | Win |
| wireline | WL |
| wireline coring | WLC |
| wireline test | WLT |
| wireline total depth | WLTD |
| wiring | WRG |
| wiring diagram | WD |
| with | w/ |
| without | W/O |
| Wolfcamp | Wolfc |
| Wolfe City | WC |
| Woodbine | WB |
| Woodford | Wdfd, Woodf |
| Woodside | Wood |
| work in place | WIP |
| work order | WO |

| | |
|---|---|
| worked | wkd |
| working | wkg |
| working interest | Wl |
| working pressure | WP |
| workover | WO, wko |
| workover rig | wkor |
| worldscale | WS |
| wrapper | wpr |
| Wreford | Wref |
| wrought iron | Wl |

## X

| | |
|---|---|
| X-ray | X-R |

## Y

| | |
|---|---|
| yard (s) | yd |
| Yates | y |
| Yazoo | Yz |
| year | yr |
| yellow | yel |
| yellow indicating lamp | YIL |
| yield point | YP |
| Yoakum | Yoak |
| your message of date | YMD |
| your message yesterdays | YMY |

## Z

| | |
|---|---|
| zenith | zen |
| Zilpha | Zil |
| zinc | ZN |
| zone | Zn |

# Abbreviations for Logging Tools and Services

The appropriate companies and associations have not established standard abbreviations for the logging segment of the oil and gas industry. The following lists by individual companies are supplemented by a Miscellaneous Section, for your convenience.

## Baker Atlas Wireline Services

| | |
|---|---|
| Acoustilog | AC |
| Acoustilog Caliper Gamma Ray | AC/CAL GR |
| Acoustilog Caliper Neutron | AC/CAL NEU |
| Acoustilog Caliper Gamma Ray Neutron | AC/CAL GR/NEU |
| Acoustic Cement Bond | CBL |
| Acoustic Cement Bond Gamma Ray | CBL/GR |
| Acoustic Cement Bond Neutron | CBL/NEU |
| Acoustic Cement Bond G/R Neutron | CBL/GR/NEU |
| Acoustic Signature | AC SIGN |
| BHC Acoustilog Caliper | AC/CAL |
| BHC Acoustilog Caliper Gamma Ray | AC/CAL/GR |
| BHC Acoustilog Caliper Neutron | AC/NUE |
| BHC Acoustilog Caliper G/R Neutron | AC/CAL/GR/NEU |
| BHC Acoustilog Caliper (Thru Casing) | AC/CAL |
| BHC Acoustilog Caliper Gamma Ray (Thru Casing) | AC/CAL/GR |
| BHC Acoustilog Caliper G/R Neutron (Thru Casing) | AC/CAL/GR/NEU |
| Caliper | CAL |
| Casing Potential Profile | CPP |

Standard Oil & Gas Abbreviator

| | |
|---|---|
| Chlorinlog | CHL |
| Chlorinlog-Gamma Ray | CHL GR |
| Compensated Densilog Caliper | CDL/CAL |
| Compensated Densilog Caliper Gamma Ray | CDL/CAL/GR |
| Compensated Densilog Caliper Neutron | CDL/CAL/NEU |
| Compensated Densilog Caliper G/R Neutron | CDL/CAL/GR/NEU |
| Compensated Densilog Caliper Minilog | CDL/CAL/ML |
| Conductivity Derived Porosity | CDP |
| Compensated Z-Densilog | CZDL |
| Corgun | SWC |
| Densilog Caliber Gamma Ray Log | CDL/CAL/GR |
| Depth Determination | DD |
| Directional Survey | DIR |
| Dual Induction Focused Log | DIFL |
| Dual Induction Focused Log Gamma Ray | DIFL/GR |
| Electrolog | EL |
| Formation Tester | FMT |
| 4 Arm High Resolution Diplog | DIP |
| Frac Log | DIP FRAC |
| Frac Log-Gamma Ray | DIP FRAC/GR |
| Gamma Ray Cased Hole | GR |
| Gamma Ray/Dual Caliper | GR/CALD |
| Gamma Ray-Open Hole | GR |
| Gamma Ray Neutron-Cased Hole | GR/NEU |
| Gamma Ray Neutron-Open Hole | GR/NEU |
| Induction Electrolog | IEL |
| Induction Electrolog Gamma Ray | IEL/GR |
| Induction Electrolog Neutron | IEL/NEU |
| Induction Electrolog Gamma Ray Neutron | IEL/GR/NEU |
| Induction Log | IEL |
| Induction Log-Gamma Ray | IEL/GR |
| Induction Log-Neutron | IEL/NEU |
| Laterolog | LL |
| Laterolog-Gamma Ray | LL/GR |

## Abbreviations for Logging Tools and Services

| | |
|---|---|
| Laterolog-Neutron | LL/NEU |
| Laterolog-Gamma Ray-Neutron | LL/GR/NEU |
| Microlaterolog-Caliper | MLL/CAL |
| Minilog Caliper | ML/CAL |
| Minilog Caliper Gamma Ray | ML/CAL/GR |
| Moveable Oil Pot | MOP |
| Neutron (Cased Hole) | NEU |
| Neutron (Open Hole) | NEU |
| Neutron Lifetime | NLL |
| Neutron Lifetime Gamma Ray | NLL/GR |
| Neutron Lifetime Neutron | NLL/NEU |
| Neutron Lifetime G/R-Neutron | NLL/GR/NEU |
| Nuclear Flolog | NFL |
| Nuclear Flolog-Gamma Ray | NFL/GR |
| Nuclear Flolog-Gamma Ray Neutron | NFL/GR/NEU |
| Nuclear Flolog-Neutron | NFL/NEU |
| Perforating Control | PFC |
| PFC Gamma Ray | PFC GR |
| PFC Neutron | PRC NEU |
| Photon | PHT |
| Proximity Minilog | PROX/ML |
| Segmented Bond Log | SBT |
| Sidewall Neutron | SWN |
| Sidewall Neutron-Gamma Ray | SWN/GR |
| Temperature Differential | TEMP |
| Temperature-Gamma Ray-Neutron | TEMP/GR/NEU |
| Temperature Log | TL TEMP |
| Temperature Log-Gamma Ray | TEMP/GR |
| Temperature-Neutron | TEMP/NEU |
| Total Time Integrator | TT |
| Tracer Log | TRL |
| Tracer Log-Neutron | TRL/NEU |

## Schlumberger Well Services

| | |
|---|---|
| Amplitude Logging | A-BHC |
| Bore Hole Compensated | BHC |
| BHC sonic Logging | BHC |
| BHC Sonic-Gamma Ray Logging | BHC-GR |
| BHC Sonic-Variable Density | BHC-VD |
| Bridge Plug Service | BP |
| Borehole Televiewer | TVT |
| Caliper Logging | CAL |
| Casing Cutter Service | SCE-CC |
| Cement Bond Logging | CBL |
| Cement Bond-Gamma Ray Logging | CBL-GR |
| Cement Bond-Gamma Ray Neutron | CBL-GRN |
| Cement Bond-Neutron | CBL-N |
| Cement Bond-Variable Density Logging | CBL-VD |
| Cement Dump Bailer Service | DB |
| Computer Processed Interpretation | MCT |
| Continuous Directional Survey | CDR |
| Continuous Flowmeter | CFM, PFM |
| Customer Instrument Service | ICS |
| Data Transmission | TRD |
| Density Log | DENL |
| Depth Determinations | DD |
| Diamond Core Slicer | SS |
| Dipmeter | DIPM |
| Dipmeter-Digital | HDT-D |
| Directional Survey | DS |
| Dual Induction-Laterologging | DIL |
| Electric Logging | ES |
| Formation Density Logging | FDC |
| Formation Density-Gamma Ray Logging | FDC-GR |

## Abbreviations for Logging Tools and Services

| | |
|---|---|
| Formation Testing | FT |
| Gamma Ray Logging | GR |
| Gamma Ray-Neutron Logging | GRN |
| Gamma Ray-Sonic Logging | GRS |
| Gradiomanometer | GM |
| High-Resolution Thermometer | HRT |
| Induction-Electron Logging | I-ES |
| Induction-Gamma Ray Logging | I-GR |
| Junk Catcher | JB |
| Log Overlays | OL |
| Magnetic Taping | TPG |
| Microlog | ML |
| Neutron Logging | NL |
| Orienting Perforating Service | OPR |
| Perforating-Ceramic DPC | SCE |
| Perforating-Depth Control | PDC |
| Perforating-Expendable Shaped Charge | SCE |
| Perforating-Hyper Jet | SCH |
| Perforating-Hyper Scallop | SPH |
| Pressure Control | PC |
| Production Combination Tool Logging | PCT |
| Production Packer Service | PPS |
| Proximity-Microlog | ML |
| Radioactive Tracer logging | RTP |
| Rwa Logging | FAL |
| Salt Dome Profiling | ES-ULS |
| Schlumberger | Schl. |
| Seismic Reference Service | SRS |
| Sidewall Coring | CST |
| SNP Neutron Logging | SNP |
| SNP Neutron-Gamma Ray Logging | SNP-GR |
| Synergetic Log Systems | MCT |
| Temperature Logging | T |
| Temperature-Gamma Ray Logging | T-GR |

| | |
|---|---|
| Thermal Decay Logging | TDT |
| Thru-Tubing Caliper | C-C |
| Tubing, Cutter Service | SEC-CC |
| Variable Density Logging | BHC-VD |
| Variable Density-Gamma Ray Logging | VD-GR |

## Halliburton Log Service (HAL)

| | |
|---|---|
| Analog Computer Service | An Cpt. Ser |
| Caliper | Cal |
| Compensated Acoustic Velocity Log | Com AVL |
| Compensated Acoustic Velocity Log-Gamma Ray | Com AVL-G |
| Compensated Acoustic Velocity Log-Neutron | Com AVL-N |
| Compensated Density | Com Den |
| Compensated Density Gamma Ray | Com Den-GR |
| Computer Analyzed Log | CAL |
| Contact Caliper | Cont |
| Continuous Drift | Con Dr. |
| Density | Den |
| Density Gamma Ray | Den-G |
| Depth Setermination | DeDet |
| Digital Tape Recording | Dgt Tp Rec |
| Dip Log Digital Log | Dgt Dip Rec |
| Drift | Dr |
| Drill Pipe Electric Log | DPL |
| Electric Log | EL |
| Electro-Magnetic Corrosion Detector | Cor Det |
| Fluid Travel Log | FTrL |
| Formation Tester | FT |
| FoRxo Caliper | FoRxo |
| Frac-Finder Micro-Seismogram | FF-MSG |
| Frac-Finder Micro-Seismogram Gamma | FF-MSG-G |
| Frac-Finder Micro-Seismogram Neutron | FF-MSG-N |
| Gamma Guard | G-Grd |
| Gamma Ray | GR |
| Gamma Ray Depth Control | GRDC |
| Guard | Grd |
| High Temperature Equipment | HETq |

| | |
|---|---|
| Induction Electric Gamma Ray | IEL-G |
| Induction Electric Neutron | IEL-N |
| Induction Electric | IEL |
| Induction Gamma Ray | Ind-G |
| Induction Gamma | Ind-G |
| Micro-Seismogram Log, Cased Hole | MSG-CBL |
| Micro-Seismogram Gamma Collar Log, Cased | MSG-CBL-G |
| Micro-Seismogram Neutron Collar Log, Cased | MSG-CBL-N |
| Neutron Log | NL |
| Neutron Depth Control | NDC |
| Precision Temperature | Pr Temp |
| Radiation Guard | R/A Grd |
| Radioactive Tracer | R/A Tra |
| Resistivity Dip | Dip |
| Sidewall Coring | SWC |
| Sidewall Neutron | SWN |
| Sidewall Neutron-Gamma Ray | SWN-G |
| Simultaneous Gamma Ray-Neutron | GRN |
| Special Instrument Service | Sp Inst Ser |
| True Vertical Depth | TVD |

Abbreviations for Logging Tools and Services

## Log Heading Nomenclature

| Name | Abbrev. | Tool Type |
|---|---|---|
| Acoustic Caliber Log | ACE | |
| Aluminum Clay Log | | ACT |
| Adaptive Electromagnetic Propagation Log | ADEPT | EPT-G |
| Array Sonic Log | | SDT |
| Automatic Diverter Flowmeter | ADF | ADF |
| Auxiliary Measurements Log | AMS | AMS |
| Borehole Compensated Sonic Log | BHC | SLT |
| Borehole Geometry Log | BGL | BGT |
| Bridge Plus | BP | RST |
| Casing Collar Log | CCL | CAL, CCL |
| Cement Bond-Variable Density Log | CBL-VDL | SLT, SDT, CBT |
| Cement Evaluation Log | CET | CET |
| Compensated Neutron Log | CNL | CNT-A/H |
| Corrosion And Protection Evaluation Log | CPEL, CPET | CPET |
| Cyberdip | CYDIP | |
| Cyberlock | CYL | |
| Cyberscan | CYTDT | |
| Deep Propagation Log | DPL | DPL |
| Depth Determination Log | DD | |
| Downhole Seismic Array | DSA | DSA-A |
| Dual Dipmeter | SHDT | SHDT |
| Dual-Burst Thermal Neutron Log | TDT-P | TDT-P |
| Dual-Energy Neutron Log | DNL | CNT-G |
| Dual Induction SFL Log | DIL | DIT |
| Dual Laterolog | DLL | DLT |
| Dump Bailer | DB | DB |
| Electromagnetic Propagation Time Log | EPT | EPT |

297

## Standard Oil & Gas Abbreviator

| | | |
|---|---|---|
| Electromagnetic Thickness Log | ETT | ETT-A |
| Enerjet Perforating | EEJ | EEJ |
| Formation Density | FDC | PGT |
| Formation Microscanner | FMS | MEST |
| Formation Microimager | FMI | FMI |
| Four-Arm Caliber Log | CAL | BGT, HDT, SHDT |
| Fracture Identification Log | FIL | HDT |
| Gamma Ray Log | GR | SGT |
| Gamma Ray Spectroscopy Log | GST | GST |
| Geochemical Log | GLT | GST, NGT, CNT, ACT |
| Guidance Continuous Log | GCT | GCT |
| Hyperdome Perforating | HD | HD |
| Hyperjet Perforating | HJ, HJII, HJIII | HJ, HJII, HJIII |
| H2S Cable | HS CBL | |
| Induction-Spherically Focused Log | ISF | IRT, DIT |
| Inclinometer Survey, Directional | IS | GPIT, SHDT, FMS, PMI, HDT |
| Junk Basket | JB | JB |
| Litho-Density Log | LDL | LDT |
| Litho-Density Quicklook | CYLDT | |
| Long-Spacing Sonic Log | LSS | SLT, SDT |
| Mechanical Sidewalk Coring or Rotary Cores | MSCT | MSCT |
| Microlog | ML | MPT, MLT, PCD |
| Microlaterolog | MLL | MPT |
| Microlaterolog Focused Log | MSFL | SRT |
| Multifinger Caliper Log | MFC | MFCT |
| Multifrequency Electromagnetic Thickness Log | METL | ETT-D |
| Multiple Isotope Tracer | MTT | |
| Natural Gamma Ray Spectroscopy Log | NGS | NGS |

## Abbreviations for Logging Tools and Services

| | | |
|---|---|---|
| Oil Base Mud Dipmeter | OBDT | OBDT |
| Perforation Depth Control Log | PDC | CAL, CCL |
| Phasor Induction Log | PIL | DIT-E |
| Pipe Analysis Log | PAL | PAT |
| Pivotgun | PG | PG |
| Posiset Thru-Tubing Mechanical Plug-Back | MPBT | |
| Pressure Control | PC | PC |
| Production Logging Quicklook | CYPL | |
| Repeat Formation Tester Log | RFT | RFT |
| Repeat Formation Tester Quicklook | CYRFT, RFQL | |
| Reservoir Saturation Tool | RST | |
| Seismic Quicklook | CYWST | |
| Sidewall Coring | CST | CST |
| Simultaneous Production Log | PLT | PLT |
| Stratigraphic High-Resolution Dipmeter Log | SHDT | SHDT |
| Temperature Log | HRT | HTT |
| Thermal Neutron Decay Time Long | TDT | TDT |
| Triaxial Seismic Survey | SAS | SAT |
| Ultrajet Perforating | UJ | UJ |
| Ultrapack Perforating | UP | UP |
| Well Head Equipment (Pressure Control) | WHE-A,B, C,D,H,BM | |
| Well Seismic Survey | WST | WST |

## Miscellaneous

| | |
|---|---|
| Acoustic Amplitude | AAL |
| Acoustic Cement Bond G/R Neutron | CBL GRN |
| Acoustic Cement Bond Neutron | CBL N |
| Acoustic Cement Bond | CBL |
| Acoustic Fracture Identification | AFI |
| Acoustic or Acoustilog | SL |
| Acoustic Parameter | ACP |
| Acoustic Parameter-Depth | AC PAR D |
| Acoustic Parameter-Logging | AC PAR L |
| Acoustic Parameter-16mm Scope | AC PAR 16 |
| Acoustic Scope Picture | ASL |
| Acoustic Signature | AC SIGN |
| Acoustic Velocity | AVL |
| Acoustilog | ALC |
| Acoustilog Caliper-Gamma Ray | ALC-GR |
| Acoustilog Caliper-Gamma Ray-Neutron | ALC-GRN |
| Acoustilog Caliper-Neutron | ALC-N |
| After Pay Out | APO |
| Amplitude | AMP |
| Amplitude Sonic | ASL |
| Area of Mutual Interest | AMI |
| Atlantic Chlorinlog | A CHL |
| Audio Logging | AUD |
| BHC Acoustilog Caliper | BHC ALC |
| BHC Acoustilog Caliper (Thru Casing) | BHC AL TC |
| BHC Acoustilog Caliper G/R Neutron | BHC ALC GRN |
| BHC Acoustilog Caliper G/R Neutron (Thru Casing) | BHC AL GRN TC |
| BHC Acoustilog Caliper Gamma Ray | BHC ALC GR |
| BHC Acoustilog Caliper Neutron | BHC ALC N |

## Abbreviations for Logging Tools and Services

| | |
|---|---|
| Barrel | BBL |
| Before Pay Out | BPO |
| Borehole Audio Tracer Survey | BATS |
| Borehole Compensated | BHC |
| Borehole Compensated Sonic | BHCS |
| Borehole Geometry Log | BGT |
| Borehole Televiewer | BTL |
| Bottom Hole Contribution | BHC |
| Bulk Density | BLKD |
| Caliper | CALP, CL |
| Caliper Analysis | CALA |
| Caliper Curve | CALC |
| Carbon-Oxygen | C O |
| Casing Collar | CCL |
| Casing Inspection/Electro-Magnetic-Detector | CI |
| Casing Potential Profile | CPP |
| Cement Bond | CBND |
| Cement Evaluation | CEL |
| Cement Top Location | CTL |
| Cemotop | CTL |
| Channel Survey | CHNL |
| Channelmaster | CML |
| Channelmaster-Neutron | CML N |
| Chloride | CL |
| Chloride Detection | CLDL |
| Chlorinlog-Gamma Ray | CHL GR |
| Collar/Collar Correlation | PDS |
| Compensated Acoustic Velocity | CAVL |
| Compensated Densilog Caliper | CDCL |
| Compensated Densilog Caliper G/R Neutron | CDLC GRN |
| Compensated Densilog Caliper Gamma Ray | CDCL GR |
| Compensated Densilog Caliper Minilog | CDLC M |
| Compensated Densilog Caliper Neutron | CDLC N |
| Compensated Density Caliper | CDC |

| | |
|---|---|
| Compensated Density Log | CDL |
| Compensated Formation Density | CFD |
| Compensated Formation Density Caliper | CFDC |
| Compensated Gamma | CG |
| Compensated Neutron Density | CNDL |
| Compensated Neutron Log | CNL |
| Compensated Neutron Log Porosity | CNLP |
| Dual Induction Gamma Log | DIGL |
| Dual Induction Lateral/Dual Induction Focus | DILL |
| Dual Induction Dual Induction Log | DIL |
| Dual Induction SFL | DIL |
| Dual Induction Spherically Focused | DISF |
| Dual Laterolog/Microspherically Focused | DLL/MSFL |
| Dual Laterolog | DLL |
| Dual Porosity Compensated Neutron | DNL |
| Dual Resistivity Induction Log | DRI |
| Dual Sand | DUSD |
| Dual Spacing Log | DSL |
| Electrical | ES |
| Electrolog | EL |
| Electromagnetic Propagation | EPT |
| Epithermal Neutron | ETN |
| Experimental | E, XPTL |
| Farm In | FI |
| Farm Out | FO |
| Flowmeter | FLO |
| Fluid Travel | FLO |
| Fluid Travel Log/Fluid Entry Survey | FLTR |
| Focus | FOCL |
| Focused Diplog | F DIP |
| Formation Analysis | FAL |
| Formation Density | FD |
| Formation Density Caliper | FDC |
| Formation Factor | FMF |

## Abbreviations for Logging Tools and Services

| | |
|---|---|
| Formation Tester | FT |
| Forxo | MLL |
| Four Dimensional | 4D |
| 4 Arm High Resolution Diplog | R H DIP |
| Frac Log | FRAC L |
| Frac Log-Gamma Ray Log | FRAC-GR |
| Fracture Finder | ASL |
| Fracture Finder/Failure ID | FF |
| Fracture Identification Log | FIL |
| Full Bore Flowmeter | FB FM |
| Gamma Compensated Density | GCD |
| Gamma Gamma | GG |
| Gamma Gamma Density | GGD |
| Gamma Gamma Log | GGL |
| Gamma Guard | GCRD |
| Gamma Guard EL | GGRD |
| Gamma Ray | GR |
| Gamma Ray—Multi-Spaced Neutron Log | CR-MSN |
| Gamma Ray—Neutron | GRNL |
| Gamma Ray—Neutron Log | GR-N |
| Gamma Ray—Sonic | GRSL |
| Gamma Ray—Tracer Survey | GRTS |
| Gamma Ray Cased Hole | GR CH |
| Gamma Ray Depth Control | GRDC |
| Gamma Ray Neutron | GRN |
| Gamma Ray Neutron-Cased Hold | GRN CH |
| Gamma Ray Neutron-Open Hole | GR/N OH |
| Gamma Ray-Open Hole | G/R OH |
| Gamma Ray/Dual Caliper | GR/D CALIPER |
| Gamma Spectrometry | GST |
| Gas Detection | GASD |
| Gas Purchase Agreement | GPA |
| Geophone | GEO |
| Gradiomanometer | GRMR |

| | |
|---|---|
| Gravity | GRAV |
| Guard | GRDL |
| Gyro Survey | GYRO |
| Held by Production | HBP |
| High Resolution Dipmeter | HRD |
| Hydrocarbon or Gas Detection | GT |
| Hydrocarbon or Gas Detection | HCDS |
| Inclination | INCL |
| Induction | IL |
| Induction-Electric Log | I-EL |
| Induction-Lateral | ILL |
| Induction-Laterolog | I-LL |
| Induction Electrolog | IEL |
| Induction Electrolog-Gamma Ray | IEL-GR |
| Induction Electrolog-Gamma Ray Neutron | IEL-GRN |
| Induction Electrolog-Neutron | IEL-N |
| Induction Spherically Focused | ISF |
| Induction-Electric or Induction-Electro | IES |
| Isotron | I, ISOL |
| Joint Operating Agreement | JOA |
| Joint Venture Agreement | JVA |
| Joint Venture Drilling | JVD |
| Land Owner Royalty | LOR |
| Laser Log | LASR |
| Lateral | LATL |
| Laterlog-Gamma Ray-Neutron Log | LL GR-N |
| Lifetime Log | LL |
| Limestone Log | LSL |
| Limestone Device | LI |
| Liquid Isotope Injector | LII |
| Liquefied Natural Gas | LNG |
| Liquefied Petroleum Gas | LPG |
| Litho-Density | LDT |
| Lithology | LITH |

## Abbreviations for Logging Tools and Services

| | |
|---|---|
| Logger's Total Depth | LTD |
| Long-Spaced Sonic | LSS |
| Lost Circulation | LS |
| Manometer | MAN |
| Measurement for Natural Gas | MCF |
| Microlog | ML, MICL |
| Microlateral | MLAT |
| Microseismogram | MSMG |
| Microsonic Gamma Ray | MSG |
| Microspherically Focused Log | MSFL |
| Microsurvey | MS, MICS |
| Mini | ML, MINL |
| Minifocus | MLL, MINF |
| Mobile Picture | MP |
| Mono Electric | MONO |
| Mud Log/Focus Log | MUD |
| Multi-Shot Survey | MSS |
| Natural Gamma Ray Spectroscopy | NGS |
| Net Revenue Interest | NRI |
| Neutron | N, NEUT |
| Neutron Collar Log | NCL |
| Neutron Formation Density | NFD |
| Neutron Lifetime | NLL |
| Notice of Acquisition | NOA |
| Nuclear | NUCR |
| Nuclear Flow | FLO |
| Nuclear Magnetism | NML, NMAGL |
| Overriding Royalty | ORR |
| Overriding Royalty Interest | ORI |
| Perforated Log | PERF |
| Perforated Depth Control | PDC |
| Perforating Formation Collar | PFC |
| Permalog | PL |
| Permeability Spinner Survey | PSS, PRMS |

Standard Oil & Gas Abbreviator

| | |
|---|---|
| Photo | PHOT |
| Photoclinometer | PHCL |
| Pipe Analysis Log | PAL |
| Pipe Recovery Log | PRL |
| Porosity | POR |
| Proximity | PROX |
| Proximity Log | PROXL |
| Proximity-Microlog | PML |
| Quitclaim | QC |
| Radioactive Tracer | RAT, RTRS |
| Refracture | "REFR |
| Repeat Formation Tester | RFT |
| Resistivity | RES |
| Resistivity Water Apparent | RWA |
| Right of Way | ROW |
| Salinity | CL |
| Saraband | "SBND |
| Scattered Gamma Ray | CTL |
| Scope Picture Analysis | SPA |
| Section Gauge | CAL |
| Seismic Reference Survey/ Neutron | SRSN |
| Seismic Velocity Survey | SVS |
| Shear Amplitude | SA |
| Sidewall Frac Log | SFL |
| Sidewall Neutron | SN |
| Sidewall Neutron Porosity | SNP |
| Sidewall Sampler | CST |
| Sonic Caliper Log | SCL |
| Sonic Log | SL, SONL |
| Sonic Seismogram | SSMG |
| Spectral | SPCT |
| Spherical | SPH |
| Spontaneous Potential | SP |

## Abbreviations for Logging Tools and Services

| | |
|---|---|
| Strata | STRTS-tructual Exploration |
| Structural Exploration | SE |
| Synergetic | SYGT |
| Televiewer | TV |
| Temperature | HRT |
| Temperature Difference Log | TDL |
| Temperature Survey Log | TMPL |
| Thermal Decay Time | TDT |
| Three Dimensional | 3D |
| Time Log | TIME |
| Tracer Survey | TRCR |
| Two Dimensional | 2D |
| Ultralong Spacing Electric Log | ULSEL |
| Uranium | URAN |
| Variable Density | VD |
| Velocity | VEL |
| Velocity Seismic Profile | VSP |
| Velocity Survey | VRS |
| Velocity Survey Profile | VSP |
| Viscosity | VISC |
| Water Location Survey | WLS |
| Wave Form Digitizing | BHC-WFD |
| Wave Form Logging | BHC-WFL |
| Well Seismic | WST |
| Working Interest | WI |
| X-Y Caliper | XYCL |

# Federal Environmental Acronyms

Compliments of BNA Plus

| | |
|---|---|
| AHERA | Asbestos Hazard Emergency Response Act of 1986 (Title II of TSCA) |
| ANSI | American National Standards Institute |
| ATSDR | Agency for Tonic Substances and Disease Registry |
| AWT | Advanced Wastewater Treatment |
| BACT | Best Available Control Technology |
| BAT | Best Available Technology |
| BCT | Best Conventional Pollutant Control Technology |
| BOD | Biologic Oxygen Demand |
| BPT | Best Practicable Control Technology |
| BTU | British Thermal Unit |
| CAA | Clean Air Act |
| CAIR | Comprehensive Assessment Information Rule (under TSCA) |
| CAS | Chemical Abstract Service |
| CEPP | Chemical Emergency Preparedness Program |
| CERCLA | Comprehensive Environmental Response, Compensation, and Liability Act (The Superfund Law) |
| CFR | Code of Federal Regulations |
| CPSC | Consumer Product Safety Commission |
| CWA | Clean Water Act |
| CZM | Coastal Zone Management |
| DOT-E | Designation Of Material Exempt from Department of Transportation regulations |
| EP | Extraction Procedure |
| EPA | Environmental Protection Agency |

# Federal Environmental Acronyms

| | |
|---|---|
| EPCRA | Emergency Planning and Community Right-To-Know Act (Title III of SARA, commonly called Right-To-Know or SARA Title III) |
| ERC | Emissions Reduction Credit |
| FIFRA | Federal Insecticide, Fungicide, and Rodenticide Act |
| HCS | OSHA Hazard Communication Standard (Worker Right-To-Know) |
| HMTA | Hazardous Materials Transportation Act |
| HSWA | Hazardous and solid Waste Amendments (1984 RCRA Amendments |
| IARC | International Agency for Research on Cancer |
| ITC | Interagency Testing Committee |
| LAER | Lowest Achievable Emission Rate |
| LUST | Leaking Underground Storage Tanks |
| MSDS | Material Safety Data Sheet |
| MTB | Material Transportation Bureau of the Department of Transportation |
| NESHAP | National Emission Standard for Hazardous Air Pollutants |
| NIOSH | National Institute for Occupational Safety and Health |
| NOAA | National Oceanic and Atmospheric Administration |
| NPDES | National Pollutant Discharge Elimination System |
| NPRM | Notice Proposed Rulemaking |
| NRC | Nuclear Regulatory Commission or National Response Center |
| NSPS | New Source Performance Standards |
| NTP | National Toxicology Program |
| ORM | Other Regulated Materials |
| OSHA | Occupational Safety and Health Administration |
| OSHRC | Occupational Safety and Health Review Commission |
| OTA | Office of Technology Assessment |
| PEL | Permissible Exposure Limit |
| pH | Potential of Hydrogen: a measure of acidity and alkalinity |

Standard Oil & Gas Abbreviator

| | |
|---|---|
| PM | 10-micron Particulate Matter |
| PMN | Premanufacture Notification (under TSCA) |
| POTWs | Publicity Owned Treatment Works |
| PRPs | Potentially Responsible Parties |
| PSD | Prevention Significant Deterioration |
| psia | Pounds per Square Inch Absolute |
| RACT | Reasonable Available Control Technology |
| RCRA | Resource Conservation and Recovery Act |
| RQ | Reportable Quantity |
| RSPA | Research and Special Programs Administration of the Department of Transportation |
| RTECS | Registry of Toxic Effects of Chemical Substances |
| SARA | Superfund Amendments and Reauthorization Act of 1987 |
| SDWA | Safe Drinking Water Act |
| SIP | State Implementation Plan |
| SNUR | Significant New Use Rule |
| SOCMI | Synthetic Organic Chemical Industry |
| SPCC PLAN | Spill Prevention Control and Countermeasure Plan |
| SQG | Small Quantity Generation (of Hazardous waste) |
| SRF | State-administered water pollution control Revolving Funds |
| SWDA | Solid Waste Disposal Act |
| TSCA | Toxic Substances Control Act |
| TSS | Total Suspended Solids (non-filterable) |
| UIC | Underground Injection Control |
| UN | United Nations |
| UPAC | International Union Of Pure and Applied Chemistry |
| USDW | Underground Source of Drinking Water |
| UST | Underground Storage Tank |
| VHAP | Volatile Hazardous Air Pollutant |
| VOC | Volatile Organic Compound |
| Z list | OSHA list of hazardous chemicals (29 CFR 1910 Subpart Z, Worker Right-To-Know) |

## Pipe Coating Terminology and Definitions

| | |
|---|---|
| anode | Corrosion prevention device |
| C.P. | Cathodic protection or Corrosion Protection |
| dope | Pipe coating |
| dresser | Mechanical coupling used to join joints or lengths or pipe rather than threading or welding |
| FB | Flat-bottom mill or shoes |
| granny rag | Type of coating or method of coating a pipeline in the field rather than factory applied coating |
| holiday | Hole in the protective coating of a steel pipeline in the field |
| hot spot | Corrosive area located along the length of a pipeline; usually a wet bog, marsh, or bentonite area |
| I.P. | Intermediate pressure pipeline |
| IWRC | Independent wire rope center |
| jeep | Same as Holiday |
| jeeper | Electronic device or instrument used to detect holes (holidays) in the steel pipeline protective coating |
| overbend | High spot in a pipeline usually installed by field-bending a pipeline joint |
| P/C or P/W | "Painted and coated" or "painted and wrapped" pipelines-steel pipe with protective external coating of one of several different types |
| pig | Pipeline cleaning and measuring tool |
| pit catcher | Used to remove pipeline pig |
| pig launcher | Used to insert pipeline pig |
| stub | Length of small-diameter distribution pipeline form the main line to the customer's property or meter location |

| | |
|---|---|
| thin film | Type of epoxy coating for pipeline coating rather than threading or welding |
| tube turn | Prefabricated piece of pipeline (allows change of direction of a pipeline without field bending the pipeline) |
| sag | Low spot in a pipeline usually installed by field bending a pipeline joint |
| WB | Wavy bottom mill or shoes |
| wrap | Same as dope; protective coating on a steel pipeline |
| XTC | Extra-coat protective pipeline coating made of polyethylene or polypropylene material |

# Mnemonics
## Service Names

| Service Mnemonic | Service |
|---|---|
| 2CAL | 2-Arm Caliper Log |
| 2CEX | 2-Arm Caliper Log (Extended Reach) |
| 3CAL | 3-Arm Caliper Log |
| 4CAL | 4-Arm Caliper Log |
| 4CEX | 4-Arm Caliper Log (Extended Reach) |
| AC | BHC Acoustilog |
| ACL | Long-Spaced BHC Acoustilog |
| BAL | Bound Attenuation Log |
| BHJ | Bottom-Hole Junk Shot |
| BO | String Shot—Back Off |
| BO1 | Tubing |
| BO2 | Drill Pipe, Wash Pipe, Casing, Liners |
| BO3 | Drill Collars |
| BP | Bridge Plug |
| CAC | Circumferential Acoustilog |
| CBIL | Circumferential Borehole Imaging Log |
| CBL | Acoustic Cement Bond Log |
| CC | Chemical Cutters |
| CC1 | 1-2-1/16-in. (25–52mm) Pipe |
| CC2 | 2-3/8-4-in. (60–102mm) Pipe |
| CC3 | 4-1/4-5-9-/16 in. (108–141 mm) Pipe |
| CCL | Casing Collar Locator |
| CDL | Compensated Densilog |
| CHFT | Cased Hole Formation Tester |
| CIS | Customer Instrument Service |
| CN | Compensated Neutron Log |
| CPP | Casing Potential Profile |

| | |
|---|---|
| CRET | Cement Retainer |
| CST | Customer Steering Tool |
| DAC | Digital Array Acousticlog sm |
| DB | Dump Bailer |
| DD | Depth Determination |
| DEL2 | Dielectric Log—200 MHz |
| DEL4 | Dielectric Log—47 MHz |
| DGR | Digital Gamma Ray Log |
| DIEL | Dielectric Log |
| DIFL | Dual Induction-Focused Log |
| DIP | Dual Induction-Focused Log High Resolution (4-Arm) Diplog® |
| DLL | Dual Laterolog |
| DMAG | Digital Magnelog |
| DMGL | Digital Magneline |
| DPIL | Dual Phase Induction Log |
| DRB | Dual Receiver Bond Log |
| DSL | Dig ital Spectralog |
| DVL1 | Digital Vertiline 4-1/2 in. (114 mm) |
| DVL2 | Digital Vertiline 5-1/2 in. (140 mm) |
| DVL3 | Digital Vertiline 7 in. (178 mm) |
| DVL4 | Digital Vertiline 8-5/8 8 in. (219 mm) |
| DVRT | Digital Vertilog |
| DVT1 | Digital Vertilog 4-1/2 in (114 mm) |
| DVT2 | Digital Vertilog 5-1/2 in. (140 mm) |
| DVT3 | Digital Vertilog 7 in. (178 mm) |
| DVT4 | Digital Vertilog 8-5/8 8 on. (219 mm) |
| DVTL | Digital Vertiline |
| DWP | Downhole Wireline Packoff |
| EMO | Electromagnetic Orientation Tool |
| EMT | Electromagnetic Fishing Tool |
| FCON | Fluid Conductivity Log |
| FDBM | Buoyancy Measurement Fluid Density Log |
| FDDP | Differential Pressure Fluid Density Log |

## Mnemonics

| | |
|---|---|
| FDN | Nuclear Fluid Density Log |
| FG | Feeler Gauge |
| FMBK | Basket Flowmeter |
| FMCS | Continuous Spinner Flowmeter |
| FMFI | Folding Impeller Flowmeter |
| FMT | Formation Multi-Tester |
| FPM | Magnetic Freepoint Indicator |
| FPMT | Magna-Tector® Freepoint Indicator |
| FPST | Spring-TectorTM Freepoint Indicator |
| FPTM | Tri-Mag Freepoint Indicator |
| FQPG | Fast Response Quartz Pressure Gauge |
| GR | Gamma Ray |
| GUN | Guns |
| HCS | Hydraulic Cleanout Service |
| HDIP | Hexdip |
| HYDL | Hydrologsm |
| IEL | Induction Electrolog |
| JCGR | Junk Catcher Gauge Log |
| JCS | Jet Cutters |
| JCS1 | Jet Cutters 1-2-7/8-in (25–73 mm) Pipe |
| JCS2 | Jet Cutters 3-5-1/2-in (76–140 mm) Pipe |
| JCS3 | Jet Cutters 6-7-5/8-in (152–194 mm) Pipe |
| LL3 | Laterolog |
| MAG | Magnelog |
| MFC | Multi-Finger Caliper |
| MGLN | Magneline |
| ML | Minilog® |
| MLL | Micro Laterolog |
| MSI | MSI Carbon / Oxygen Log |
| MST | Metal Severing Tool |
| MVL1 | Multi-Channel Vertiline 4-1/2 in. (114 mm) |
| MVL2 | Multi-Channel Vertiline 5-1/2 in. (140 mm) |
| MVL3 | Multi-Channel Vertiline 7 in. (178 mm) |
| MVL4 | Multi-Channel Vertiline 8-5/8 in. (219 mm) |

| | |
|---|---|
| MVRT | Multi-Channel Vertilog |
| MVT1 | Multi-Channel Vertilog 4-1/2 in. (114 mm) |
| MVT2 | Multi-Channel Vertilog 5-1/2 in. (140 mm) |
| MVT3 | Multi-Channel Vertilog 7 in. (178 mm) |
| MVT4 | Multi-Channel Vertilog 8-5/8 in. (219 mm) |
| MVTL | Multi-Channel Vertiline |
| NEU | Neutron Log |
| NFL | Nuclear Flolog |
| ORIT | Instrument Orientation Log |
| PDB | Positive Displacement Dump Bailer |
| PDK | PDK-100® Log |
| PFC | PFC Gamma Ray Log |
| PFN | PFC Neutron Log |
| PHT | Photon Log |
| POL | Perforating Orientation Log |
| PPKR | Production Packer |
| PPL | Polymer Pathfinder Log |
| PRL | Pipe Recovery Log |
| PROX | Proximity Log |
| PRSM | PRISM® Log |
| RCOR | Rotary Coring Tool |
| QPG | Quartz Pressure Gauge |
| SB | Sinker Bar |
| SBT | Segmented Bond Tool |
| SL | Spectralog® |
| SON | Sonan Log |
| SPG | Strain Pressure Gauge |
| SRB | Single Receiver Bond Log |
| SRPL | Surface Recorded Pressure Log |
| SSH | Surface Shot |
| ST | FMT Break Off (PVT) Sample Tanks |
| STHS | FMT Break Off (PVT) Sample Tanks ($H_2S$) |
| SWAT | Swing-Arm Tracer Log |
| SWC | Sidewall Corgun |

## Mnemonics

| | |
|---|---|
| SWN | Sidewall Epithermal Neutron Log |
| TBFS | Through-Tubing Borehole Fluid Sampler |
| TBRT | Thin-Bed Resistivity |
| TCAL | Through-Tubing (X-Y) Caliper Log |
| TEMP | Differential Temperature Log |
| TFLR | TTRM Sub Fluid Resistivity |
| TRL | Tracerlog |
| TTBP | Through-Tubing Bridge Plug |
| TTEM | TTRM Sub Temperature |
| TTEN | TTRM Sub Tension |
| TTRM | TTRM Sub |
| UDIP | Ultrasonic Diplog |
| VIBR | Vibrator |
| VL1 | Vertiline 3-1/2 in. (89 mm, 4-1/2 in. (114 mm) |
| VL2 | Vertiline 5-1/2 in. (140 mm) |
| VL3 | Vertiline 7 in. (178 mm) |
| VL4 | Vertiline 8-5/8 in. (219 mm) |
| VPC | Formation Multi-Tester |
| VRT | Vertilog® |
| VRT1 | Vertilog 3-1/2 in. (89 mm), 4-1/2 in. (114 mm) |
| VRT2 | Vertilog 5-1/2 in. (140 mm) |
| VRT3 | Vertilog 7 in. (178 mm) |
| VRT4 | Vertilog 8-5/8 in. (219 mm) |
| VTLN | Vertiline |
| WHI | Water Holdup Indicator |
| ZDL | Compensated Z-Densilog$^{SM}$ |

Standard Oil & Gas Abbreviator

## Computational Products

Product
Mnemonic  Computational Service

### Open Hole Analysis
|  | HORIZON |
| --- | --- |
| HCE | Category Estimation |
| HME | Magnitude Estimation |
| HLS | Layer/Square |
| HDI | Category Data Import |
| HCI | Core Data Import |
| OPTM | OPTIMA® |
| LIN | LINEAR |
| TBA | Thin-Bed Analysis |
| CLAS | Clay Analysis and Shaly Sand Evaluation (CLASS®) |
| CLAY | Clay Description (CLAYS) |
| CRA | Complex Reservoir Analysis |
| DIE | Dielectric Analysis |
| RMA | Radioactive Mineral Analysis |
| FI | Fracture Index |
| SA | Sandstone Analysis |
| CS | Coal Seam Analysis |
| STRA | STRATA LOGIK® |

### Acoustic Waveform Processing
| DDM | DDBHC Multi-Shot Processing |
| --- | --- |
| DDS | DDBHC Slowness Waveform Analysis |
| MSP | Multi-Shot Processing |
| SWA | Slowness Waveform Analysis |
| EWA | Energy Waveform Analysis |
| WPT | Waveform processing, DDBHC$\Delta$T |

|      | Rock Properties |
|------|------|
| MP   | Mechanical Properties |
| SS   | Sand Strength |
| FP   | Fracture Migration |
| DARC | Permeability Estimation |

### Image Processing

|      |      |
|------|------|
| ACCR | Accelerometer Corrections |
|      | Preprocessing (CIBLSM PREP) |
| RSA  | Resample |
| NSE  | Noise Reduction |
| ORI  | Orientation |
|      | Borehole Corrections |
| MUD  | Mud Attenuation |
| ECC  | Tool Eccentricity |
|      | Image Enhancement |
| EDG  | Edge Detection |
| SIG  | Sigma Smoothing |
| THLD | Threshholding |
| HGM  | Histogram Equalization |

### Interactive Image Processing

|      |      |
|------|------|
| SDC  | Synthetic Dip Curves (1-4) |
| SDCA | Additional Dip Curves |
| OC   | Other Curves |
| SNP  | Snap Shot |
| SNPA | Additional Copies |
| XSEC | Cross Section |
| BOUT | Breakout |
| VIEW | View |
| CLRP | Color Plot Continuous |
| RAY  | Optimized Grey Scales |
| RAYP | Additional Playback |
| IREP | Integrated Report |

### Diplog Analysis
STRATAGON℠
| | |
|---|---|
| SOP | Optical |
| SAT | Automatic |
| SDIP | STRATA DIP® |
| DIPC | Basic Diplog Computation |
| AZD | Azimuth Dipfrac |
| DS | Directional Survey |
| TBT | True Bed Thickness |
| TVT | True Vertical Thickness |
| TVD | True Vertical Depth |

### Cased Hole Analysis
PDK-100® Log Analysis (SEARCH)
| | |
|---|---|
| STL | Time Lapse |
| LIL | Log-Inject-Log |
| SOH | With Openhole Data Included |
| SCH | Basic |

MSI C/O Log Analysis (CHES II®)
| | |
|---|---|
| CTL | Time Lapse |
| COH | With Openhole Data Included |
| CHES | Basic |
| PLFO | PROFLOW |
| PRM | PRISM® |
| GAS | Gas Storage |

Pipe Evaluation
| | |
|---|---|
| VTRC | Vertilog/Vertline |
| CPPC | CPP |

### Pressure Transient Analysis
| | |
|---|---|
| LAYR | Multi-Layer Testing |
| PAN | Pan System Multi-Rate Test |
| PTA | Pressure Transient Analysis (REALITY) |
| PDR | Pressure Data Report |
| NODE | Well Performance Simulation (NODES) |
| FMTC | FMT Analysis |

# Abbreviations for Companies, Associations, & Organizations

## U.S. and Canada

| | |
|---|---|
| AADE | American Association of Drilling Engineers |
| AAODC | See IADC |
| AAPG | American Association of Petroleum Geologists |
| AAPL | American Association of Professional Landmen |
| AAR | Association of American Railroads |
| ABSORB | Alaska Beaufort Seat Oil Spill Response Body |
| ACMP | Alaska Costal Management Program |
| ACS | American Chemical Society |
| ADDC | Association of Desk and Derrick Clubs |
| ADSC | The International Association of Foundation Drilling |
| AEC | Atomic Energy Commission |
| AECRB | Alberta Energy Conservation Resources Board |
| AEG | Association of Engineering Geologists |
| AESP | Association of Energy Services Companies |
| AGA | American Gas Association |
| AGI | American Geological Institute |
| AGTL | Alberta Gas Trunkline Co., LTD. |
| AGU | American Geophysical Union |
| AIChE | American Institute of Chemical Engineers |
| AIME | American Institute of Mining, Metallurgical and Petroleum Engineers |
| AISI | American Iron and Steel Institute |
| ALCOA | Aluminum Company of America |
| ANSI | American National Standards Institute |
| AOAC | Association of Official Agricultural Chemists |

| | |
|---|---|
| AOCS | American Oil Chemists Society |
| AOGA | Alaskan Oil & Gas Association |
| AOPL | Association of Oil Pipe Lines |
| AOSC | Association of Oilwell Servicing Contractors |
| AP&VMA | American Paint & Varnish Manufacturers Assoc. |
| APHA | American Public Health Association |
| API | American Petroleum Institute |
| APRA | American Petroleum Refiners Association |
| APW | Association of Petroleum Writers |
| ARCO | Atlantic Richfield Co. |
| ARKLA | Arkansas Louisiana Gas Co. |
| ASA | American Standards Assoc. |
| ASCE | American Society of Civil Engineers |
| ASHRAE | American Society of Heating, Refrigerating, and Air-Conditioning Engineers, Inc. |
| ASLE | American Society of Lubricating Engineers |
| ASME | American Society of Mechanical Engineers |
| ASPG | American Society of Professional Geologists |
| ASPOE | American Society of Petroleum Operations Engineers |
| ASSE | American Society of Safety Engineers |
| ASSMR | American Society for Surface Mining and Reclamation |
| ASTM | American Society for Testing Materials |
| AWS | American Welding Society |
| AWWA | American Water Works Association |
| BLM | Bureau of Land Management |
| BLS | Bureau of Labor Statistics |
| BP | British Petroleum |
| BuMInes | Bureau of Mines, U.S. Department of the Interior |
| CADE | Canadian Association of Drilling Engineers |
| CAGC | A combine: Continental Oil Co., Atlantic Richfield Co., and Cities Service Oil Co. |
| CAGCT | Centre for Advanced Gas Combustion Technology |

# Abbreviations for Companies, Associations, & Organizations

| | |
|---|---|
| CAODS | Canadian Association of Oilwell Drilling Contractors |
| CAPP | Canadian Association of Petroleum Producers |
| CCCOP | Conservation Committee of California Oil Producers |
| CDS | Canadian Development Corp. |
| CEPA | Canadian Energy Pipeline Association |
| CFRC | Coordinating Fuel Research Committee |
| CGA | Canadian Gas Association |
| CGA | Clean Gulf Associates |
| CGRI | Canadian Gas Research Institute |
| CGTC | Columbia Gas Transmission Co. |
| CIM | Canadian Institute of Mining, Metallurgy and Petroleum |
| CIPA | California Independent Petroleum Association |
| CL&F | Continental Land and Fur Co. |
| CNG | Consolidated Natural Gas Supply Company |
| CNGP | CNG Producing Committee |
| COE | Corps of Engineers |
| COGA | Colorado Oil & Gas Association |
| CONOCO | Continental Oil Co. |
| COPAS | Council of Petroleum Accountants Society |
| CORCO | Commonwealth Oil Refining Co., Inc. |
| CORS | Canadian Operational Research Society |
| CPA | Canadian Petroleum Association |
| CPIA | Cathodic Protection Industry Association |
| CRC | Coordinating Research Council, Inc. |
| CSEG | Canadian Society of Exploration Geophysicist |
| CSPG | Canadian Society of Petroleum Geologists |
| D&D | Desk & Derrick |
| DNR | Department of Natural Resources |
| DOE | Department of Energy |
| DOT | Department of Transportation |
| DRAPR | Delaware River Area Petroleum Refineries |
| Drssr., da | Dresser Atlas |

| | |
|---|---|
| EMR | Department of Energy, Mines, and Resources (Canada) |
| ENTELEC | Energy Telecommunications & Electrical Association |
| EPA | Environmental Protection Agency |
| ERCB | Energy Resource Conservation Board (Alberta, Canada) |
| FAA | Federal Aviation Agency |
| FCC | Federal Communications Commission |
| FERC | Federal Energy Regulatory Commission |
| FPC | Federal Power Commission |
| FSIWA | Federation of Sewage and Industrial Wastes Assoc. |
| FTC | Federal Trade Commission |
| FTPI | Fiberglass Tank & Pipe Institute |
| GAMA | Gas Appliance Manufacturers Association |
| GE | General Electric Company |
| GISB | Gas Industries Standards Board |
| GM | General Motors |
| GNEC | General Nuclear Engineering Co. |
| GR&DC | Gulf Research and Development Company |
| GREG | Groupe de recherche en gazotechnologies |
| GTC | Gas Technology Canada |
| IADC | International Association of Drilling Contractors (formerly AAODC) |
| IAE | Institute of Automotive Engineers |
| IAGC | International Association of Geophysical Contractors |
| ICC | Interstate Commerce Commission |
| IEEE | Institute of Electrical and Electronics Engineers |
| IGT | Institute of Gas Technology |
| IGUA | Industrial Gas Users Association |
| ILMA | Independent Lubricant Manufacturers Association |
| ILTA | Independent Liquid Terminals Association |
| INGAA | Independent Natural Gas Association of America |

## Abbreviations for Companies, Associations, & Organizations

| | |
|---|---|
| IOCA | Independent Oil Compounders Association |
| IOCC | Interstate Oil Compact Commission |
| IOGA | Illinois Oil & Gas Association |
| IOGANY | Independent Oil & Gas Association of New York |
| IOGAP | Independent Oil & Gas Association of Pennsylvania |
| IOGAWY | Independent Oil & Gas Association of West Virginia |
| IOGCC | Interstate Oil & Gas Compact Commission |
| IOPA | Independent Oil Producers Agency |
| IOSA | International Oil Scouts Association |
| IP | Institute of Petroleum |
| IPAA | Independent Petroleum Association of America |
| IPAC | Independent Petroleum Association of Canada |
| IPCA | International Petroleum Credit Association |
| IPE | International Petroleum Exposition |
| IPP/L | Interprovincial Pipe Line Co. |
| IRAA | Independent Refiners Association of America |
| ISA | Instrument Society of America |
| ISSA | International Slurry Surfacing Association |
| IUPIW | International Union of Petroleum & Industrial Workers |
| JCUMWE | Joint Committee on Uniformity of Methods of Water Examination |
| KERMAC | Kerr-McGee Corp. |
| KIOGA | Kansas Independent Oil and Gas Association |
| KOGA | Kentucky Oil & Gas Association |
| LIOGA | Louisiana Independent Oil & Gas Association |
| LL&E | Louisiana Land & Exploration Co. |
| MAPL | Michigan Association of Petroleum Landmen |
| MBGS | Michigan Basin Geological Society |
| MDEG | Michigan Department of Environmental Quality |
| MDNR | Michigan Department of Natural Resources |
| MIOP | Mandatory Oil Import Program |
| MMS | Minerals Management Service |
| MOGA | Michigan Oil & Gas Association |

| | |
|---|---|
| MPA | Michigan Petroleum Association |
| MPSC | Michigan Public Service Commission |
| NACD | National Association of Chemical Distributors |
| NACE | National Association of Corrosion |
| NACOPS | National Advisory Committee on Petroleum Statistics (Canada) |
| NACR | National Association of Chemical Recyclers |
| NAEP | National Association of Environmental Professionals |
| NAESCO | National Association of Energy Services Companies |
| NAOHSM | National Association of Oil Heating Service Managers |
| NARO | National Association of Royalty Owners |
| NAS | National Academy of Science |
| NASA | National Aeronautical and Space Administration |
| NCPSA | Natural Gas Processors Suppliers Association |
| NDA | National Drilling Association |
| NEB | National Energy Board (Canada) |
| NEMA | National Electrical Manufacturers Association |
| NEPA | National Environmental Policy Act of 1969 |
| NGPA | Natural Gas Processor Association |
| NGSA | Natural Gas Supply Association |
| NLGI | National Lubricating Grease Institute |
| NLPGA | National Liquefied Petroleum Gas Association |
| NLRB | National Labor Relations Board |
| NMA | National Mining Association |
| NMOCC | New Mexico Oil Conservation Commission |
| NMOGA | New Mexico Oil and Gas Association |
| NOFI | National Oil Fuel Institute |
| NOIA | National Ocean Industries Association |
| NOJC | National Oil Jobbers Council |
| NOMADS | National Oil-Equipment Manufacturers and Delegates Society |
| NPC | National Petroleum Council |

## Abbreviations for Companies, Associations, & Organizations

| | |
|---|---|
| NPDES | National Pollution Discharge Elimination System |
| NPGA | National Propane Gas Association |
| NPR | Naval Petroleum Reserve |
| NPRA | Naval Petroleum Reserve, Alaska |
| NPRA | National Petroleum Refiners Association |
| NPTO | National Petroleum Technology Office |
| NRC | Nuclear Regulatory Commission |
| NSF | National Science Foundation |
| NTOGA | North Texas Oil & Gas Association |
| OCAW | Oil, Chemical & Atomic Workers International Union |
| OCR | Office of Coal Research |
| OEP | Office of Emergency Preparedness |
| OERB | Oklahoma Energy Resources Board |
| OFA | Oxygenated Fuels Association |
| OGJ | Oil & Gas Journal |
| OIA | Oil Import Administration |
| OIAB | Oil Import Appeals Board |
| OIC | Oil Information Committee |
| OIPA | Oklahoma Independent Petroleum Association |
| OOC | Offshore Operators Committee |
| OOGA | Ohio Oil & Gas Association |
| OPC | Oil Policy Committee |
| OPI | Ontario Petroleum Institute |
| ORIAC | Oil Refining Industry Action Committee |
| ORSANCO | Ohio River Valley Water Sanitation Commission |
| OXY | Occidental Petroleum Corp. |
| PAD | Petroleum Administration for Defense |
| PCF | Petroleum Communications Foundation |
| PEI | Petroleum Equipment Institute |
| PESA | Petroleum Equipment Suppliers Association |
| PETCO | Petroleum Corporation of Texas |
| PETSOC | Petroleum Society of the Canadian Institute of Mining, Metallurgy and Petroleum |

| | |
|---|---|
| PGAC | Propane Gas Association of Canada |
| PGCOA | Pennsylvania Grade Crude Oil Association |
| PIEA | Petroleum Industry Electrical Association |
| PITS | Petroleum Industry Training Service |
| PLATO | Pennzoil Offshore Gas Operators |
| PLCA | Pipe Line Contractors Association |
| PMAA | Petroleum Marketers Association of America |
| POGO | Pennzoil Offshore gas Operators |
| PPC | Petroleum Packaging Council |
| PPI | Plastic Pipe Institute |
| PPROA | Panhandle Producers and Royalty Owners Association |
| PRI | Petroleum Recovery Institute |
| PSAC | Petroleum Services Association of Canada |
| PTTC | Petroleum Technology Transfer Council |
| RMOGA | Rocky Mountain Oil and Gas Association |
| R-PAT | Regional Petroleum Association |
| RPI | Research Planning Institute |
| RRC | Railroad Commission (Texas) |
| RTL | Refinery Technology Laboratory |
| SACROC | Scurry Area Canyon Reef Operators Committee |
| SAE | Society of Automotive Engineers |
| Schl., Sj | Schlumberger |
| SEG | Society of Exploration Geophysicists |
| SEPAC | Small Explorers and Producers Association of Canada |
| SEPM | Society of Economic Paleontologists and Mineralogists |
| SFER | Santa Fe Energy Resources, Inc. |
| SGA | Southern Gas Association |
| SLAM | A combine: Signal Oil and Gas Co., Louisiana Land & Exploration Co., and Marathon Oil Co. |
| SMA | Society of Mineral Analysts |
| SMENET | Society for Mining, Metallurgy & Exploration |

# Abbreviations for Companies, Associations, & Organizations

| | |
|---|---|
| SMRI | Solution Mining Research Institute |
| SOCAL | Standard Oil Company of California |
| SOHIO | Standard Oil Co. of Ohio |
| SPE | Society of Petroleum Engineers of AIME |
| SPEE | Society of Petroleum Evaluation Engineers |
| SPWLA | Society of Professional Well Log Analysts |
| SSDA | Service Station Dealers of America & Allied Trade |
| STATCAN | Statistics Canada ex Dominion, Bureau of Statistics (DBS) |
| TAPS | Trans-Alaska Pipeline Systems |
| TCP | Trans-Canada Pipe Lines Ltd. |
| TEMA | Tubular Exchange Manufacturers Association |
| TETCO | Texas Eastern Transmission Corp. |
| TGT | Tennessee Gas Transmission Co. |
| THUMS | A combine: Texaco, Inc., Humble Oil & Refining Co., Union Oil Co. of California, Mobil Oil Corp., and Shell Oil Co. |
| TIPRO | Texas Independent Producers and Royalty Owners Association |
| TRANSCO | Transcontinental Gas Pipe Line Corp. |
| TT&T | Texaco Trading & Transportation |
| UNOCAL | Union Oil of California |
| UOCO | Union Oil Company |
| UOP | Universal Oil Products Company |
| USEPA | U.S. Environmental Protection Agency |
| USGS | United States Geological Survey |
| USP | United States Pharmocopoeia |
| WeCTOGA | West Central Texas Oil and Gas Association |
| WIPA | Wyoming Independent Producers Association |
| WOGA | Western Oil & Gas Association |
| WPC | World Petroleum Congress |
| WSPA | Western States Petroleum Association |
| Wx | Welex |

Standard Oil & Gas Abbreviator

## Outside U.S. and Canada

| | | |
|---|---|---|
| AAP | Afro-Arab Petroleum Trading & Petrochemicals WLL | *Kuwait* |
| ABOI | Assoc. of British Oceanic Industries | *UK* |
| ACNA | Aziende Colori Nazionali Affini | *Italy* |
| ACP | Association des Consultants Petroliers | *France* |
| ACPA | Alexandria Co. for Petroleum Additives | *Egypt* |
| ACPHA | Association Cooperative pour la Recherche et l'Exploration des Hydrocarbures en Algerie | *Algeria* |
| ADB | Asian Development Bank | — |
| ADCO | Abu Dhabi Co. for Onshore Oil Operations | *Abu Dhabi* |
| ADDCAP | Abu Dhabi Drilling Chemicals and Products Ltd. | *Abu Dhabi* |
| ADGAS | Abu Dhabi Gas Liquefaction Co. Ltd. | *Abu Dhabi* |
| ADMA-OPCO | Abu Dhabi Marine Operating Co. | *Abu Dhabi* |
| ADNATCO | Abu Dhabi National Tanker Co. | *Abu Dhabi* |
| ADNOC | Abu Dhabi National Oil Company | *Abu Dhabi* |
| ADNOC-FOD | Abu Dhabi National Oil Company for Distribution | *Abu Dhabi* |
| ADPC | Abu Dhabi Petroleum Co. Ltd. | *Abu Dhabi* |
| ADPPOC | Abu Dhabi Petroleum Ports Operating Co. | *Abu Dhabi* |
| AEGPL | Assoc. Europeenne des Gaz de Petrole Liquifies | *France* |
| AFG | Assoc. Francaise du Gaz | *France* |
| AFPC | Al Forat Petroleum Company | *Syria and others* |
| AFSSA | Aviation Fueling Services S.A. | *Greece* |
| AFTP | Assoc. Francaise des Techniciens et Professionals du Petrole | *France* |

## Abbreviations for Companies, Associations, & Organizations

| | | |
|---|---|---|
| AGA | Australian Gas Assoc. | *Australia* |
| AGAS | Associated Gas Supplies Ltd. | *UK* |
| AGIBA | Agiba Petroleum Company | *Egypt* |
| AGIP | Azienda Generale Italiana Petroli S.p.A. (see ENI) | *Italy* |
| AGOC | Aramco Gulf Operations Co. Ltd. | *Saudi Arabia* |
| AGOCO | Arabian Gulf Oil Company | *Libya* |
| AHL | Amerada Hess Ltd. | *UK* |
| AHN | Amerada Hess Norge A/S | *Norway* |
| AIG | Australian Institute of Geoscientists | *Australia* |
| AIP | Australian Institute of Petroleum | *Australia* |
| AIPK | American International Petroleum Kazakhstan | *Kazakhstan* |
| ALBPETROL | Albanian Petroleum Corporation | *Albania* |
| ALEPCO | Ste. Algero-Arab-Libyenne d'Exploitation et de Production des Produits Petroliers | *Algeria* |
| AMI | Associazione Mineraria Italiana | Italy |
| AMPOL | Australian Petroleum Pty. Ltd. | Australia |
| AMPTC | Arab Maritime Petroleum Transport Co. | *OAPEC* |
| ANCAP | Administracion Nacional de Combustibles, Alcohol y Portland | *Uruguay* |
| ANPC | Afghan National Petroleum Co. | *Afghanistan* |
| AOC | Arabian Oil Company Ltd. | *Japan* |
| AOC | Aramco Overseas Company | *Netherlands, China, Korea* |
| APC | Alexandria Petroleum Company | *Egypt* |
| APEC | Asia Pacific Economic Cooperation Forum | — |
| APG | Assoc. of Policy Geomorphologists | *Poland* |
| API | Anonima Petroli Italiana S.p.A. | *Italy* |
| APIA | Australian Pipeline Industry Assoc. Inc. | *Australia* |
| APPEA | Australian Petroleum Production & Exploration Assoc. | *Australia* |

| | | |
|---|---|---|
| APRC | Arab Petroleum Research Center | *France* |
| APRC | Amerya Petroleum Refining Company | *Egypt* |
| APSCO | Arab Petroleum Services Co. | *OAPEC* |
| APTI | Arab Petroleum Training Institute | *OAPEC* |
| ARAMCO | (see Saudi Aramco) (formerly Arabian American Oil Company) | *Saudi Arabia* |
| ARPEL | Assistencia Reciproca Petrolera Empresarial Latinoamerica (Regional Assoc. of Oil & Natural Gas Cos. in Latin America and the Caribbean) | — |
| ASEAN | Association of Southeast Asian Nations | — |
| ASEG | Australian Society of Exploration Geophysicists | *Australia* |
| ASESA | Asfaltes Espanoles S.A. | *Spain* |
| ASRY | Arab Shipbuilding & Repair Yard Co. | *OAPEC* |
| ATAS | Anadolu Tastiyehanesi A.S. | *Turkey* |
| ATLANTICO | Compania Espanola de Petroleos Atlantico S.A. | *Spain* |
| AXORC | Assuit Oil Refining Company | *Egypt* |
| BANAGAS | Bahrain National Gas Co. BSC | *Bahrain* |
| BANOCO | Bahrain National Oil Company | *Bahrain* |
| BAPCO | Bahrain Petroleum Company BSC | *Bahrain* |
| BAPETCO | Badr El Deen Petroleum Co. | *Egypt* |
| BAPEX | Bangladesh Petroleum Exploration Company Ltd. | *Bangladesh* |
| BASF | Badische Anilin & Soda Frabrik AG | *Germany* |
| BERL | Boral Energy Resources Ltd. | *Australia* |
| BG | British Gas plc | *UK* |
| BG E&P | BG Exploration and Production Ltd. | *UK* |
| BGC | British Gas Council | *UK* |
| BHP | Broken Hill Proprietary plc | *Australia* |
| BITOR | Bitumenes Orinoco S.A. | *Venezuela* |
| BMP | Bureau of Minerals and Petroleum | *Denmark* |

## Abbreviations for Companies, Associations, & Organizations

| | | |
|---|---|---|
| BNOCL | Barbados National Oil Company Ltd. | *Barbados* |
| BNPE | Bureau de Normalisation du Petrole | *France* |
| BOMINFLOT | KG Bominflot Bunkergas fuer Mineraloele mbH & Co. | *Germany* |
| BOPEC | Bonaire Petroleum Corporation | *Neth. Antilles* |
| BORCO | Bahamas Oil Refining Co. | *Venezuela* |
| BOTAS | BOTAS, Petroleum Pipeline Corp. Ltd. | *Turkey* |
| BP | British Petroleum Company plc | *UK* |
| BPC | Bharat Petroleum Corp. | *India* |
| BR | Petrobras Internacional S.A. | *Brazil* |
| BRASPETRO | Petrobras Internacional S.A. | *Brazil* |
| BRGG | Bureau de Recherches Geologiques et Geophysique | *France* |
| BRGM | Bureau de Recherches Geologiques et Minieres | *France* |
| BSI | British Standards Institute | *UK* |
| BTL | Bantry Terminals Ltd. | *Ireland* |
| CAL | Caltex Australia Ltd. | *Australia* |
| CALTEX | Various affiliates of Texaco Inc, and Chevron Corp. partnership | — |
| CAMEL | Companie Algerienne du Methane Liquide | *France/Algeria* |
| CAMPSA | Compania Arrendataria del Monopolio de Petroleos, S.A. (sub. of Repsol) | *Spain* |
| CAPAG | Entreprise Moderne de Canalisations Petrolieres, Aquiferes et Gasieres | *France* |
| CARBESA | Carbon Black Espanola S.A. | *Spain* |
| CCC | Clean Caribbean Cooperative | — |
| CEDIPSA | Compania Espanola Distribuidora de Petroleos S.A. | *Spain* |
| CEP&M | Comite d'Etudes Petrolieres et Marine | *France* |
| CEPCO | City Ecuadoriana Production Co. | *Ecuador* |

| | | |
|---|---|---|
| CEPE | Corporacion Estatal de Petroleo Ecuatoriano | *Ecuador* |
| CEPSA | Compania Espanola de Petroleos, S.A. | *Spain* |
| CERHYD | Centre de Recherche Pour La Valorisation des Hydrocarbures & Derives | *Algeria* |
| CFG | Companie Francaise pour le Developpement de Geothermie et des Energies Nouvelles | *France* |
| CFM | Companie Francaise du Methane | *France* |
| CFP | Companie Francaise du Petroles (also Total) | *France* |
| CFR | Companie Francaise de Raffinage (sub. of Total) | *France* |
| CGES | Centre for Global Energy Studies | *UK* |
| CGG | Companie Generale de Geophysique | *France* |
| CHR | Surgutneftegas | *Russia* |
| CIEPSA | Compania de Investigacion y Exploraciones Petroliferas, S.A. | *Spain* |
| CIM | Companie Industrielle Maritime | *France* |
| CINSA | Compania Insular del Nitrogena, S.A. | *Spain* |
| CIRES | Companhia Industrial de Resinas Sinteticas | *Portugal* |
| CKTA | Central Kazakhstan Territorial Administration for Natural Resource Protection | *Kazakhstan* |
| CLH | Compania Logistica de Hidrocarburos S.A. | *Spain* |
| CMPT | Centre for Marine & Petroleum Technology | *UK* |
| CNACG | China National Administration of Coal Geology | *China* |
| CNI | Conoco Norway | *Norway* |
| CNODC | China National Oil & Gas Exploration & Development Corp. (see CSPC) | *China* |

## Abbreviations for Companies, Associations, & Organizations

| | | |
|---|---|---|
| CNOOC | China National Offshore Oil Corporation | *China* |
| CNPC | China National Petroleum Corporation | *China* |
| CNSI | Conoco North Sea Inc. | *Norway* |
| COCHIME | Companie Chimique de la Mediterranee | *France* |
| CODESSAS | Cia. Operadora de Estaciones de Servicio S.A. de CV | *Mexico* |
| COLPET | Colombian Petroleum Co. | *Colombia* |
| COMESA | Cie. Mexicana de Exploraciones S.A. | *Mexico* |
| COMGAS | Companhia de Gas de Sao Paulo | *Brazil* |
| COMMEGO | Committee on Geology & Mineral Resources | *Bulgaria* |
| CONCAWE | Oil companies' European organization for environment, health and safety | *Belgium* |
| CONSPAIN | Continental Oil Co. of Spain (Conoco Espanola, S.A.) | *Spain* |
| COOP | Societe Cooperative Des Petroles | *Egypt* |
| COPAREX | Companie de Participations, de Recherches et d'Exploitations Petrolieres | *France* |
| CORC | Cairo Oil Refining Company | *Egypt* |
| COSMO | | *Japan* |
| COVA | Netherlands National Petroleum Stockpiling Agency (Stichting Centraal Orgaan Voorraadvorming Aardolieprodukten) | *Netherlands* |
| CPC | Chinese Petroleum Corporation, Taiwan | *Republic of China* |
| CPC | Caspian Pipeline Consortium | *Kazakhstan* |
| CPC | Ceylon Petroleum Corporation | *Sri Lanka* |
| CPCL | Chennai Petroleum Corp. Ltd. | *India* |
| CPDP | Comite Professionel du Petrole | *France* |
| CPI | PT Caltex Pacific Indonesia | *Indonesia* |
| CRC | Czech Refinery Company | *Czech Rep.* |
| CRE | Comision Reguladora de Energia | *Mexico* |

| | | |
|---|---|---|
| CRP | Bureau de Recherche de Petrole | *France* |
| CRR | Companie Rhenane de Raffinage | *France* |
| CSPC | China Southern Petroleum Exploration & Development Corp. (formerly CNODC, sub of CNPC) | *China* |
| CUPET | Commercial Cupet S.A. | *Cuba* |
| DANA | Dana Petroleum plc | *UK* |
| DANOP | Dansk Operatorselskab | *Denmark* |
| DDGAS | South Transdanubian Gas Co. | *Germany* |
| DEA | Deutsche Erdol Aktiengesselschaft (see REW-DEA) | *Germany* |
| DELEK | Delek the Israel Fuel Corp. Ltd. | *Israel* |
| DEMINEX | DEMINEX GmbH | *Germany* |
| DEP | Public Petroleum Corp. S.A. | *Greece* |
| DEPA | Greek State Natural Gas Company | *Greece* |
| DERPROSA | Derivados del Propileno S.A. | *Spain* |
| DEUDAN | Deutsche/Danische Erdgastransport GmbH | *Germany/Denmark* |
| DEUTAG | Deutsche Tiefbohr-AG | *Germany* |
| DGMK | German Society for Petroleum & Coal Science & Technology (Deutsche Wissenschaftliche Gesellschaft fur Erdol, Erdgas und Kohle e.V.) | *Germany* |
| DICOL | Distribuidora de Combustiveis e Lubrificantes da Guinea Bissau Lda. | *Portugal* |
| DIGANAMEX | Distribuidora de Gas Natural del Estado de Mexico S.A. | *Mexico* |
| DIGAQRO | Distribuidora de Gas de Queretaro S.A. | *Mexico* |
| DIMAH | Direction des Matieres Premiers et des Hydrocarbures | *France* |
| DIOCO | DEMINEX Egypt Co. | *Egypt* |
| DISPESA | Compania Distribuidora de Productos Petroliferos S.A. | *Spain* |

## Abbreviations for Companies, Associations, & Organizations

| | | |
|---|---|---|
| DITAS | TPAO Marine Tanker Subsidiary | *Turkey* |
| DMGM | Direction Nationale de la Geologie et des Mines | *Mali* |
| DNO | Det Norske Oljeselskap A/S | *Norway* |
| DNV | Det Norske Veritas | *Norway* |
| DONG | Dansk Olie og Naturgas A/S | *Denmark* |
| DTI | Department of Trade & Industry | *UK* |
| DUC | Dansk Underground Consortium | *Denmark* |
| DUPETCO | Dubai Petroleum Co. | *Dubai* |
| E&P FORUM | Oil Industry Intenational Exploration & Production Forum | *UK* |
| EAGE | European Assoc. of Geoscientists & Engineers | *Netherlands* |
| EC | European Commission | — |
| ECN | Netherlands Energy Research Foundation (Energieonderzoek Centrum Nederland) | *Netherlands* |
| ECOPETROL | Empresa Colombiana de Petroleos | *Colombia* |
| EEC | Energy Equity Corp. Ltd. | *Australia* |
| EEG | Erdgas Erdol Gommern | *Germany* |
| EEMUA | Engineering Equipment & Materials Users Assoc. | *UK* |
| EGPC | Egyptian General Petroleum Corp. | *Egypt* |
| EGSMA | Egyptian Geological Survey & Mining Authority | *Egypt* |
| EGTA | Etudes et Grands Travaux de l'Atlantique | *France* |
| EIGS | Ethopian Institute for General Surveys | *Ethiopia* |
| EIVAL | Sociedad de Empreendimentos, Investimentos e Armazenagem de Gases S.A. | *Portugal* |
| EKTA | East Kazakhstan Territorial Administration for Natural Resource Protection | *Kazakhstan* |
| ELF | Elf Aquitaine (now TotalFinaElf) | *France* |

| | | |
|---|---|---|
| ELF NORGE | Elf Petroleum Norge A/S | *Norway* |
| ELGI | Eotvos Lorand Geofizikai Intezet (Geophysical Institute of Hungary) | *Hungary* |
| EMAGEO | Entreprise Nationale de Geophysique | *Algeria* |
| EMALCO | Empresa Almacenadora de Combustibles Ltda. | *Chile* |
| EMARAT | Emirates General Petroleum Corp. | *UAE* |
| ENACOL | Empresa Nacionale de Combustiveis, SARL | *Cape Verde* |
| ENAFOR | Entreprise Nationale de Forage | *Algeria* |
| ENAGAS | Empresa Nacional del Gas S.A. | *Spain* |
| ENAP | Empresa Nacional del Petroleo | *Chile* |
| ENFERSA | Empresa Nationale de Fertiliantes S.A. | *Spain* |
| ENGEBRAS | Engenharia Especializada Brasileira, S.A. | *Brazil* |
| ENH | Empresa Nacional de Hidrocarbonetos de Moçambique | *Mozambique* |
| ENI | Ente Nazionale Idrocarburi S.p.A. (incorporates Agip) | *Italy* |
| ENIP | Entreprise Nationale d'Industrie Petrochimique | *Algeria* |
| ENPPI | Engineering for the Petroleum and Process Industries | *Egypt* |
| EOP | Edinburgh Oil & Gas plc | *UK* |
| EPC | Ethiopian Petroleum Enterprise | *Ethiopia* |
| EPC | Egyptian Petrochemicals Company | *Egypt* |
| EPSG | European Petroleum Survey Group | — |
| ERAP | Entreprise de Recherches et d'Activites Petrolieres | *France* |
| ERIP | European/Russian Council for the Petroleum Industry | — |
| ESSAF | Esso Standard Societe Anonyme Francaise | *France* |
| ESSO REP | Societe Esso de Recherches et Exploitation Petrolieres | *France* |

## Abbreviations for Companies, Associations, & Organizations

| | | |
|---|---|---|
| ETAP | Entreprise Tunisienne d'Activites Petrolieres | *Tunisia* |
| ETDE | Energy Technology Data Exchange (intl. consortium) | — |
| ETPM | Societe Entrepose G.T.M pour les Travaux Petroliers Maritimes | *France* |
| EU | European Union | — |
| EURAFREP | Societe de Recherches et d'Exploitation de Petrole | *France* |
| EURAFRAP | EURAFRAP Nederland B.V. (and others) | *Netherlands* |
| EUROGAS | European Union of the Natural Gas Industry | *Belgium* |
| EUROGIF | European Oil and Gas Innovation Forum | — |
| EUROPIA | European Petroleum Industry Assoc. | — |
| EV | Erdoel-Vereinigung (Swiss Petroleum Assoc.) | *Switzerland* |
| FERTIL | Ruwais Fertilizer Industries Ltd. | *Abu Dhabi* |
| FIGAZ | Federation de l'Industrie du Petrole | *France* |
| FINA | PetroFina S.A. (now TotalFinaElf) | *Belgium* |
| FIOC | First International Oil Company | *Kazakhstan* |
| FPB | Belgian Oil Federation | *Belgium* |
| FRANCAREP | Compagnie Franco-Afrique de Recherches Petrolieres | *France* |
| G7 | Group of Seven (leading industrialized countries) | — |
| GAIL | Gas Authority | *India* |
| GAOCMAO | Gulf Area Oil Companies Mutual Aid Organization | — |
| GASCO | Abu Dhabi Gas Industries Ltd. | *Abu Dhabi* |
| GASCO | Egyptian Natural Gas Co. | *Egypt* |
| GASUNIE | NV Nederlandse Gasunie | *Netherlands* |
| GATOIL | Subsidiary of NOC Libya | *Libya* |

| | | |
|---|---|---|
| GAZPROM | RAO Gazprom (former USSR Ministry of Gas Production) | *Russia* |
| GCC | Gulf Cooperation Council | *Middle East* |
| GDF | Gas de France | *France* |
| GDP | Gas de Portugal | *Portugal* |
| GEISO | Geisum Oil Company | *Egypt* |
| GEOPLIN | Slovenian Gas Company | *Slovenia* |
| GEP | French Oil & Gas Suppliers Council (Groupement des Entreprises Parapetrolieres et Paragazieres) | *France* |
| GFU | Gas Negotiating Board | Norway |
| GHAIP | Ghanaian-Italian Petroleum Co., Ltd. | *Ghana* |
| GIO | Gulf Investment Organization | Kuwait/GCC |
| GNPC | Ghana National Petroleum Corp. | *Ghana* |
| GNPOC | Greater Nile Petroleum Operating Co. | *Egypt* |
| GPA EUROPE | Gas Processors Assoc.—Europe | — |
| GPC | The General Petroleum Co. | *Egypt* |
| GPEP | Gabinete Para Pesquisa e Exploracao de Petroleo | *Portugal* |
| GSPC | Gujarat State Petroleum Corp. Ltd. | *India* |
| GUPCO | Gulf of Suez Petroleum Co. | *Egypt* |
| HAR | Hellenic Aspropyrgas Refinery | *Greece* |
| HELIOS | Ste. Pour la Construction et l'Exploitation d'une Usine d'Helium Liquide et d'Azote | *Algeria* |
| HISPANOIL | Hispanica de Petroleos, S.A. (see Repsol) | *Spain* |
| HOC | Hindustan Organic Chemicals Ltd. | *India* |
| HPCL | Hindustan Petroleum Corporation Ltd. | *India* |
| HYDRO | Norsk Hydro | *Norway* |
| HYDRO-CONGO | National Oil Exploration and Exploitation Corp. | *Congo* |
| HYLSA | Hojalata y Lamina, S.A. | *Mexico* |
| IAEE | International Assoc. for Energy Economics | *US* |

## Abbreviations for Companies, Associations, & Organizations

| | | |
|---|---|---|
| IAP | Institut Algerien du Petrole | *Algeria* |
| IAPG | Instituto Argentino del Petroleo y del Gas | *Argentina* |
| IBP | Instituto Brasileiro de Petroleo | *Brazil* |
| ICGTI | International Centre for Gas Technology Information | *IEA* |
| ICI | Imperial Chemical Industries Ltd. | *UK* |
| ICIANZ | Imperial Chemical Industries of Australia & New Zealand Ltd. | *UK* |
| ICPA | International Cooperative Petroleum Assoc. | *Netherlands* |
| IEA | International Energy Agency | — |
| IEEJ | Institute of Energy Economics, Japan | *Japan* |
| IEF | International Energy Forum | — |
| IEOC | International Egyptian Oil Co., Inc. | *Egypt* |
| IFC | International Finance Corporation | *World Bank* |
| IFE | Institute of Energy Technology | *Norway* |
| IFP | Institut Francais du Petrole | *France* |
| IGSAS | Istanbul Fertiliser Industry Corp. | *Turkey* |
| IGU | International Gas Union | *France* |
| IMIQ | Instituto Mexicano de Ingenieros Quimicos AC | *Mexico* |
| IMO | International Maritime Organization | — |
| INA | Industrija Nafte | *Croatia* |
| INDEIN | Ingenieria y Desarrolio Industrial S.A. | *Spain* |
| INH | Instituto Nacional de Hidrocarburos | *Spain* |
| INI | Instituto Nacional de Industria | *Spain* |
| INOC | Iraq National Oil Company | *Iraq* |
| INPC | Irish National Petroleum Corp. Ltd. | *Ireland* |
| INPEX | Indonesian Petroleum Ltd. | *Japan* |
| INPP | Institut Nationale de Plongee Professionnelle | *France* |
| INTERQUISA | Intercontinental Quimica S.A. | *Spain* |

| | | |
|---|---|---|
| INTERTANKO | Intl. Assoc. of Independent Tanker Owners | — |
| IOC | Indian Oil Company | *India* |
| IOOA | Irish Offshore Operators' Assoc. | *Ireland* |
| IOPC | Intl. Oil Pollution Compensation | — |
| IP | The Institute of Petroleum | *UK* |
| IP | Italiana Petroli S.p.A. | *Italy* |
| IPA | Indonesian Petroleum Assoc. | *Indonesia* |
| IPE | The International Petroleum Exchange of London Ltd. | *UK* |
| IPIC | International Petroleum Investment Co. | *UAE* |
| IPIECA | International Petroleum Industry Environmental Conservation Assoc. | *UK* |
| IRBP | Institut Royal Belge du Petroles | *Belgium* |
| IRO | Assoc. of Dutch Suppliers in the Oil & Gas Industry | *Netherlands* |
| ISNEFT | Israel National Oil Co. Ltd. | *Israel* |
| ISO | International Organization for Standardization | — |
| JAPEX | Japan Petroleum Exploration Co. Ltd. | *Japan* |
| JAPT | Japanese Assoc. for Petroleum Technology | *Japan* |
| JEC | Japan Energy Corporation | *Japan* |
| JGA | Japan Gas Association | *Japan* |
| JIS | Japanese Industrial Standards | *Japan* |
| JKX | JKX Oil and Gas plc | *UK* |
| JNOC | Japan National Oil Corporation | *Japan* |
| JUPC | Jubail United Petrochemical Co. | *Saudi Arabia* |
| KAFCO | Kuwait Aviation Fueling Co. KSC | *Kuwait* |
| KBB | Kavernen Bau- und Betriebs-GmbH | *Germany* |
| KCS | Kazakhstancaspishelf Joint Stock Co. | *Kazakhstan* |
| KMOC | Khanty Mansiysk Oil Corp. | *Siberia, Russia* |
| KNOC | Korean National Oil Corporation | *Korea* |
| KNPC | Kuwait National Petroleum Co. KSC | *Kuwait* |
| KOC | Kuwait Oil Company KSC | *Kuwait* |

## Abbreviations for Companies, Associations, & Organizations

| | | |
|---|---|---|
| KOGAS | Korean Gas Corp. | *Korea* |
| KOTC | Kuwait Oil Tanker Co. | *Kuwait* |
| KPC | Kuwait Petroleum Corp. | *Kuwait* |
| KPC | Khalda Petroleum Company | *Egypt/Spain* |
| KTM | Kazakhturmunay Ltd. Joint Petroleum Corp. | *Turkey/Kazakhstan* |
| KUFPEC | Kuwait Foreign Petroleum Exploration Co. KSC | *Kuwait* |
| KVO | Koninklijke Van Ommeren NV | *Netherlands* |
| LASMO | LASMO plc (London & Scottish Marine Oil Co.) | *UK* |
| LGA | Laboratoire de Geophysique Applique | *Algeria* |
| LGSC | Liquified Gas Shipping Co. | *Abu Dhabi* |
| LOR | Lindsey Oil Refinery Ltd. | *UK* |
| LUBRIESUR | Lubricantes del Sur S.A. | *Spain* |
| LUBRINER | Lubricantes Nervion S.A. | *Spain* |
| LUBRITURIA | Lubricantes Turia S.A. | *Spain* |
| LUK | Lukoil | *Russia* |
| MABANAFT | Marquard & Bahls AG | *Germany* |
| MAMR KAZGOSNEDRA | Main Administration of Mineral Resources | *Kazakhstan* |
| MARICONSULT | Technical support to Transmed | *Algeria* |
| MEDCO | Mediterranean Oil Shipping & Transport Co. SAL | *Lebanon* |
| MEGAL | Mittle-Europaische-Gasleitungsgesellschaft | *Germany* |
| MEKOG | Maatschappij Tot Exploitatie van Kooksovengassen NV | *Netherlands* |
| METG | Mittelrheinische Ergastransport GmbH | *Germany* |
| MGT | Mongol Petroleum Co. | *Mongolia* |
| MIDTAP | Middle East Oil Tankage & Pipelines | *Egypt* |

| | | |
|---|---|---|
| MIGOP | Misr Gulf Oil Processing Company | *Egypt* |
| MIRO | Mineraloelraffinerie Oberrhein GmbH & Co. | *Germany* |
| MITI | Ministry of International Trade and Industry | *Japan* |
| MND | Czech National Oil Company | *Czech Rep.* |
| MOC ACOR | Distrubuidora de Cumbustiveis S.A. | *Mozambique* |
| MOCL | Mobil Oil Co. Ltd. | *UK* |
| MODEC | Mitsui Ocean Development & Engineering Co., Ltd. | *Japan* |
| MOL | Magyar Olaj es Gasipari Reszvenytarsasag (Hungarian Oil & Gas Co. Ltd.) | *Hungary* |
| MPAL | Magellan Petroleum Australia Ltd. | *Australia* |
| MPC | Misr Petroleum Company | *Egypt* |
| MPDC | Mitsubishi Petroleum Development Co. Ltd. | *Japan* |
| MPL | Murco Petroleum Limited | *UK* |
| MRPM | Bureau de Recherches et de Participations Minieres Morocco | *Morocco* |
| MTA | General Directorate of Mineral Research & Exploration | *Turkey* |
| MWV | Assoc. of the German Petroleum Industry (Mineralolwirtschaftsverband) | *Germany* |
| NAFTAL | Entreprise Nationale de Raffinage et de Distribution des Produits Petroliers | *Algeria* |
| NAFTEC | Refining subsidiary of SONATRACH | *Algeria* |
| NAFTOGAZ | Centre de Developpement et d'Application des Techniques Petrolieres et Gaieres | *Algeria* |
| NAM | Nederlandse Aardolie Mij BV | *Netherlands* |
| NAMCOR | National Petroleum Corp. of Namibia | *Namibia* |
| NATREF | National Petroleum Refiners of South Africa (Pty) Ltd. | *South Africa* |

## Abbreviations for Companies, Associations, & Organizations

| | | |
|---|---|---|
| NDC | National Drilling Co. | *Abu Dhabi* |
| NDSM | Nederlandse Dok en Scheepsbouw Maatschappij | *Netherlands* |
| NEREFCO | Netherlands Refining Co. | *Netherlands* |
| NETG | Nordrheinische Erdgastransport GmbH | *Germany* |
| NETRA | Norddeutsche Erdgas Transversale & Co. | *Germany* |
| NGI | Norge Geotekniske Institut | *Norway* |
| NGSCO | National Shipping Co. | *Abu Dhabi* |
| NGV | Natural Gas Vehicles Company | *Egypt* |
| NIDOCO | Nile Delta Oil Company | *Egypt and others* |
| NIGC | National Iranian Gas Company | *Iran* |
| NIOC | National Iranian Oil Co. | *Iran* |
| NIPC | National Iranian Petrochemical Company | *Iran* |
| NIRPDC | National Iranian Refining & Petroleum Products Distribution Co. | *Iran* |
| NKTA | West Kazakhstan Territorial Administration for Natural Resource Protection | *Kazakhstan* |
| NLNG | Nigerian LNG | *Nigeria* |
| NMS | National Marine Service Co. | *Abu Dhabi* |
| NNPC | National Petroleum Construction Co. | *Abu Dhabi* |
| NNPC | Nigerian National Petroleum Corp. | *Nigeria* |
| NOC LIBYA | National Oil Company (Libya) | *Libya* |
| NOCK | National Oil Corporation of Kenya Ltd. | *Kenya* |
| NODCO | National Oil Distribution Co. | *Qatar* |
| NOGEPA | Netherlands Oil & Gas Exploration & Production Assoc. | *Netherlands* |
| NP | Trinidad & Tobago National Petroleum Marketing Co. Ltd. | *Trinidad & Tobago* |
| NPC | National Petroleum Company | *Barbados* |
| NPC | Nasr Petroleum Co. | *Egypt* |
| NPD | Norwegian Petroleum Directorate | *Norway* |

| | | |
|---|---|---|
| NPIC | National Petrochemical Industrialization Co. | *Saudi Arabia* |
| NWO | Nord-West Oelleitung GmbH | *Germany* |
| NZOG | New Zealand Oil and Gas Ltd. | *New Zealand* |
| OAPEC | Organization of Arab Petroleum Exporting Countries | — |
| OCENSA | Oleoducto Central S.A. | *Colombia* |
| OCIMF | Oil Companies International Marine Forum | — |
| OEA | Operaciones Especiales Argentinas | *Argentina* |
| OECD | Organization for Economic Cooperation & Development | — |
| OEP | Organization for Energy Planning | *Egypt* |
| OGDC | Oil and Gas Development Corp. | *Pakistan* |
| OGON | Onroerend Goed Oranje-Nassau NV | *Netherlands* |
| OGP | International Assoc. Of Oil & Gas Producers | *UK* |
| OILINVEST | Intl. Arm of NOC Libya | *Libya* |
| OKGT | Hungarian National Oil & Gas Trust | *Hungary* |
| OKIOC | Offshore Kazakhstan Intl. Operating Co. | *Kuwait* |
| OLADE | Organization Latinoamericana de Energia | *Ecuador* |
| OLECASA | Oleoductos Canarios S.A. | *Spain* |
| OLF | Norwegian Oil Industry Assoc. | *Norway* |
| OMNIS | Office des Mines Nationales et des Indusries Strategiques (formerly Office Militaire National Pour Les Industries Strategiques) | *Madagascar* |
| OMV | Oesterreichische Mineraloelverwaltung A.G. | *Austria* |
| ONACO | Orenburg Oil Company | *Russia* |
| ONAREP | Office National des Recherches et d'Exploitations Petrolieres (Moroccan National Oil Co.) | *Morocco* |

## Abbreviations for Companies, Associations & Organizations

| | | |
|---|---|---|
| ONE | Oranje-Nassau Energie | *Netherlands* |
| ONEPM | Oranje-Nassau Energy Participate | *Netherlands* |
| ONGC | Oil and Natural Gas Commission | *India* |
| OPAB | Oljeprospektering AB | *Sweden* |
| OPC | Oriental Petrochemicals Company | *Egypt* |
| OPEC | Organization of Petroleum Exporting Countries | — |
| OPECNA | OPEC News Agency | *OPEC* |
| OPITO | Offshore Petroleum Industry Training Organization | *UK* |
| ORL | Oil Refineries Ltd. | *Israel* |
| OSL | Oil Search Ltd. | *New Guinea* |
| OSO | Oil, Gas & Petrochemicals Supplies Office | *UK* |
| PAJ | Petroleum Assoc. of Japan | *Japan* |
| PAN ANDEAN | Pan Andean Resources plc | *UK* |
| PARCO | Pak-Arab Refinery | *Pakistan/Abu Dhabi* |
| PAS | Petroleum Air Services Company | *Egypt* |
| PC | Perez Comerc | *Argentina* |
| PCJ | Petroleum Corp. of Jamaica | *Jamaica* |
| PCK | Petrolchemie und Kraftstoffe AG Schwedt | *Germany* |
| PDB | Petronas Dagangan Bhd. | *Malaysia* |
| PDO | Petroleum Development Oman | *Oman* |
| PDV | Petroleos de Venezuela (also PDVSA) | *Venezuela* |
| PDVSA | Petroleos de Venezuela | *Venezuela* |
| PEMEX | Petroleos Mexicanos | *Mexico* |
| PEQUIVEN | Petroquimica de Venezuela | *Venezuela* |
| PERMAGO | Perforaciones Marinas del Golfo S.A. | *Mexico* |
| PERSERO | PT Perusahaan Gas Negara | *Indonesia* |
| PERTAMINA | Perusahaan Pertambangan Minyak dan Gas Bumi Negara (Indonesian State Oil and Gas Enterprise) | *Indonesia* |

| | | |
|---|---|---|
| PERUPETRO | Peru Petro S.A. | *Peru* |
| PESGB | Petroleum Exploration Society of Great Britain | *UK* |
| PETRESA | Petroquimica Espanola, S.A. | *Spain* |
| PETROBANGLA | Bangladesh Oil, Gas & Mineral Corporation | *Bangladesh* |
| PETROBEL | Belayim Petroleum Co. | *Egypt* |
| PETROBRAS | Petroleo Brasileiro, S.A. | *Brazil* |
| PETROCAN | Petroleos de Canarias | *Spain* |
| PETROCI | Societe Nationale d'Operations Petrolieres de la Cote d'Ivoire | *Cote d'Ivoire* |
| PETROCOMMERCIAL | Empresa Estatal de Commercializacion y Transporte de Petroles del Ecuador | *Ecuador* |
| PETROECUADOR | Petroleos de Ecuador | *Ecuador* |
| PETROGAB | Societe Nationale Petroliere—PETROGAB | *Gabon* |
| PETROGAL | Petroleos de Portugal | *Portugal* |
| PETROGAS | Petroleum Gas Co. | *Egypt* |
| PETROJET | Petroleum Projects and Technical Consultations Co. | *Egypt* |
| PETROGUIN | National Petroleum Co. of Guinea-Bissau (formerly PETROMINAS) | *Guinea-Bissau* |
| PETROINDUSTRIAL | Empresa Estatal de Industrializacion de Petroles del Ecuador | *Ecuador* |
| PETROJAM | Petroleum Corporation of Jamaica | *Jamaica* |
| PETROKEMYA | Arabian Petrochemical Co. | *Saudi Arabia* |
| PETROL | Petrol, A Slovene Oil Company | *Slovenia* |
| PETROM | Petrom RA | *Romania* |
| PETROMED | Petroleos del Mediterraneo | *Spain* |
| PETROMIN | General Petroleum and Mineral Organization | *Saudi Arabia* |

## Abbreviations for Companies, Associations, & Organizations

| | | |
|---|---|---|
| PETROMINAS | Empresa Nacional de Pesquisa e Exploracao Petroliferas e Mineras EP (now PETROGUIN) | *Guinea-Bissau* |
| PETROMOC | Empresa Nacional Petroles de Moçambique | *Mozambique* |
| PETRON | Petron Corporation (PNOC/Saudi Aramco partnership) | *Philippines* |
| PETRONAS | Petroliam Nasional Bhd. | *Malaysia* |
| PETRONET | Petronet India ltd. | *India* |
| PETRONIC | Empresa Nicaraguense del Petroleo | *Nicaragua* |
| PETRONOR | Petroleos del Norte, S.A. | *Spain* |
| PETROPAR | Petroleos Paraguayos | *Paraguay* |
| PETROPERU | Petroleos del Peru | *Peru* |
| PETROPRODUCTION | Empresa Estatal de Exploration y Produccion de Petroleos del Ecuador | *Ecuador* |
| PETROQUISA | Petrobras Quimica S.A. | *Brazil* |
| PETROREP | Societe Petroliere de Recherches Dans La Region Parisienne | *France* |
| PETROSEN | Societe Nationale des Petroles du Senegal | *Senegal* |
| PETROSUR | Petroleos del Sur S.A. | *Spain* |
| PETROTRIN | Petroleum Company of Trinidad & Tobago Ltd. | *Trinidad & Togago* |
| PETROVIETNAM | Vietnam Oil and Gas Company | *Vietnam* |
| PETROZAIRE | Entreprise Petroliere du Zaire | *Zaire* |
| PGNIG | Polish Oil and Gas Co. | *Poland* |
| PIP | Phillips-Imperial Petroleum Ltd. | *UK* |
| PNOC | Philippine National Oil Company | *Philippines* |
| POAS | TPAO Marketing Subsidiary | *Turkey* |
| POHOL | Private Oil Holdings Oman Ltd. | *Oman* |
| POLICOLSA | Poliolefinas Colombianas S.A. | *Colombia* |
| PORMINER | Promotora de Minmercados S.A. | *Spain* |

| | | |
|---|---|---|
| PORTGAS | Sociedad de Producao e Distribucion de Gas S.A. | *Portugal* |
| PPC | Petroleum Pipelines Company | *Egypt* |
| PPL | Pakistan Petroleum Ltd. | *Pakistan* |
| PROAS | Productos Asfalticos S.A. | *Spain* |
| PRODESA | Productos de Estireno, S.A. de C.V. | *Mexico* |
| PROPEL | Produtos de Petroleo Lda. | *Portugal* |
| PROTEXA | Construcciones Protexa, S.A. de C.V. | *Mexico* |
| PSO | Pakistan State Oil Company Ltd. | *Pakistan* |
| PTB | Belayim Petroleum Company—PETROBEL | *Egypt* |
| PTIT | Petroleum Institute of Thailand | *Thailand* |
| PTSI | PT Stanvac Indonesia | *Indonesia* |
| PTT | Petroleum Authority of Thailand | *Thailand* |
| PTTEP | PTT Exploration & Production Public Co. Ltd. | *Thailand* |
| PTTEPI | PTTEP International Ltd. | *Thailand* |
| PTTEPO | PTTEP Offshore Investment Co. Ltd. | *Thailand* |
| QATARGAS | Qatar Liquified Gas Co. (consortium) | *Qatar* |
| Q-CHEM | Qatar Chemical Corp. | *Qatar* |
| QGPC | Qatar General Petroleum Corporation | *Qatar* |
| QTHERM | Queensland Department of Energy | *Australia* |
| QUIMAR | Quimica del Mar, S.A. | *Mexico* |
| RAG | Rohol-Aufsuchungs AG | *Austria* |
| RAMCO | Ramco Energy plc | *UK* |
| RASGAS | Ras Laffan LNG Co. (consortium) | *Qatar* |
| RASHPETC | Rashed Oil Company | *Egypt and others* |
| RASIOM | Raffinerie Siciliane Olii Minerali | *Italy* |
| REGANOSA | Regasification subsidiary of SONATRACH | *Algeria* |
| REPSOL | Repsol S.A. (formerly "Hispanoil") (see Repsol-YPF) | *Spain* |
| REPSOL-YPF | Former Hispanoil (Spain) & YPF (Argentina) | *Spain* |
| RESISA | Resinas Sinteticas S.A. | *Spain* |

## Abbreviations for Companies, Associations, & Organizations

| | | |
|---|---|---|
| ROSKOMNEDRA | Ministry of Fuel & Energy for the Russian Federation | *Russia* |
| RPC | Refineria de Petroleo Concon S.A. | *Chile* |
| RPL | Reliance Petroleum Ltd. | *India* |
| RRP | NV Rotterdam-Rijn Pijpleiding Mij | *Netherlands* |
| RSO | Red Sea Oil Corporation | *Tanzania* |
| RWE-DEA | RWE(=Rheinische-Westfalische Electrzitatwerk AG)-DEA AG fur Mineraloel und Chimie | *Germany* |
| SAAGA | Sociedad Acoreana de Armazenagem de Gas S.A. | *Portugal* |
| SABIC | Saudi Arabia Basic Industries Corporation | *Saudi Arabia* |
| SADCOP, MAHRUKAT | Syrian Storing and Distributing Co., Petroleum Products | *Syria* |
| SAFAF | Saudi Petrochemical Co. | *Saudi Arabia* |
| SAGA | Saga Petroleum AS | *Norway* |
| SAGE | Scottish Area Gas Evacuation Pipeline System | *UK* |
| SAMREF | Saudi Aramco Mobil Refinery Co. Ltd. | *Saudi Arabia* |
| SAPIA | South African Petroleum Industry Assoc. | *South Africa* |
| SAR | Societe Africaine de Raffinage | *Senegal* |
| SARAS | Saras S.p.A. Raffinerie Sarde | *Italy* |
| SARPI | Ste. Algerienne pour la Reparation des Pipelines | *Algeria* |
| SARPOM | Societa Azionaria Raffinazione Olli Minerali | *Italy* |
| SASOL | South African Coal, Oil and Gas Corp. Ltd. | *South Africa* |

| | | |
|---|---|---|
| SASREF | Saudi Aramco Shell Refinery Co. | *Saudi Arabia* |
| SAUDI ARAMCO | Saudi Arabian Oil Company | *Saudi Arabia* |
| SCANRAFF | Skandinaviska Raffinaderi AB | *Sweden* |
| SCPC | Saudi Chevron Petrochemical Corp. | *Saudi Arabia* |
| SCPM | Supreme Council for Petroleum & Minerals | *Saudi Arabia* |
| SECA | Societe Europeenne des Carburants | *Belgium* |
| SECWA | State Energy Commission of Western Australia | *Australia* |
| SEDIGAS | Sociedad para el Estudio y Desarrallo de la Industria del Gas S.A. | *Spain* |
| SEREPCA | Societe Elf de Recherches et d'Exploitation des Petroles au Cameroun | *Cameroon* |
| SETG | Suddeutsche Erdgastransport GmbH | *Germany* |
| SGF | Societe Geologique de France | *France* |
| SH | Saarberg Handel GmbH | *Germany* |
| SHARQ | Eastern Petrochemical Co. | *Saudi Arabia* |
| SHELL | Royal Dutch/Shell Group | *Netherlands/UK* |
| SICP | Suez Oil Company | *Egypt and others* |
| SINOPEC | Chinese Petroleum and Chemical Corporation | *China* |
| SIPC | Saudi International Petrochemicals Co. | *Saudi Arabia* |
| SIPETROL | Sociedad Internacional Petrolera S.A. | *Chile* |
| SIR | Societe Ivoirienne de Raffinage | *Cote d'Ivoire* |
| SIRIM | Standards & Industrial Research Institute of Malaysia | *Malaysia* |
| SKTA | South Kazakhstan Territorial Administration for Natural Resource Protection | *Kazakhstan* |
| SLPRC | Sierra Leone Petroleum Refining Co. Ltd. | *Sierra Leone* |

## Abbreviations for Companies, Associations, & Organizations

| | | |
|---|---|---|
| SMR | Societe Malgache de Raffinage | *Madagascar* |
| SNH | Societe Nationale des Hydrocarbures | *Cameroon* |
| SNOC | Seychelles National Oil Co. Ltd. | *Seychelles* |
| SOC | Sirte Oil Company | *Libya* |
| SOCAR | State Oil Company of the Azerbaijan Republic | *Azerbaijan* |
| SOCDET | Sydney Oil Co. Drilling & Exploration | *Australia* |
| SOCIR | Societe Congo-Italienne de Raffinage | *Dem. Rep. of the Congo* |
| SOGARA | Societe Equatoriale de Raffinage | *Gabon* |
| SOGUIP | Societe Guineane de Petroles | *Guinea* |
| SOMALGAZ | Societe Mixte Algerienne de Gaz | *Algeria* |
| SONACOL | Sociedad Nacional de Oleoductos Ltda. | *Chile* |
| SONACPP | Societe Nationale de Commercialisation des Produits Petrolieres | *Benin* |
| SONANGOL | Sociedade Nacional de Combustiveis Sonangol | *Angola* |
| SONATRACH | Entreprise Nationale Sonatrach | *Algeria* |
| SONELGAZ | Societe Nationale d'Electricite et du Gaz | *Algeria* |
| SONIDEP | Societe Nigerienne de Produits Petroliers | *Niger* |
| SONPETROL | Sondeos Petroliferos, S.A. | *Spain* |
| SOPC | Suez Oil Processing Company | *Egypt* |
| SOPOR | Sociedad Distribuidora de Cumbustiveis S.A. | *Portugal* |
| SOQUIP | Societe Quebecoise d'Initiatives Petrolieres | *Canada* |
| SOR | Sharjah Oil Refining Co. Ltd. | *Emir. of Sharjah* |
| SOTRAL | Societe de Transport d'Arzews | *Algeria* |
| SPC | Syrian Petroleum Co. | *Syria* |
| SPC | Singapore Petroleum Company Ltd. | *Singapore* |
| SPC | Societe Cherifienne des Petroles | *Morocco* |
| SPC BVI | Sonatrach Petroleum Corp. | *British V.I.* |
| SPE AB | Svenska Petroleum Exploration AB | *Sweden* |

| | | |
|---|---|---|
| SPI | Svenska Petroleum Institutet och Oljebolagen I Sverige | *Sweden* |
| SPI | Societe Petrolifera Italiana | *Italy* |
| SPIC | Southern Petrochemical Industries Corp. Ltd. | *India* |
| SPL | Shell Pakistan ltd. | *Pakistan* |
| SPL | Saudi Petroleum Ltd. | *Saudi Arabia* |
| SPLSE | Societe du Pipe-Line Sud Europeen | *France* |
| SPOL | Saudi Petroleum Overseas Ltd. | *Saudi Arabia* |
| STAATSOLIE | Staatsolie Maatschappij Suriname NV (State Oil Co. Suriname NV) | *Suriname* |
| STATOIL | Den Norske Stats Oljeselskap AS | *Norway* |
| STEG | Societe Tunisienne d'Electricite et de Gaz | *Tunisia* |
| SUCO | Suez Oil Company | *Egypt* |
| SUMED | Arab Petroleum Pipelines Co. | *Egypt & others* |
| SUT | Society for Underwater Technology | *UK* |
| SYCOPOL | Oil Spill Control Assoc. of France | *France* |
| TAL | Deutsche Transalpine Oelleitung GmbH | *Germany* |
| TAL AUSTRIA | Transalpine Olleitungin Oesterreich GmbH | *Austria* |
| TAMOIL | Subsidiary of NOC Libya | *Libya* |
| TAMSA | Tubos de Acero de Mexico, S.A. | *Mexico* |
| TATNEFT | Tatneft | *Tatarstan, Russia* |
| TATSA | Tanques de Acero Trinity, S.A. | *Mexico* |
| TECHINT | Compania Technica Internacional | *Brazil* |
| TENP | Trans Europa Naturgas Pipeline GmbH | *Germany* |
| TEPSA | Terminales Portuadias S.A. | *Spain* |
| TEXSPAIN | Texaco (Spain) Inc. | *Spain* |
| TMPC | Transmediterranean Pipeline Co. Ltd. | *Algeria* |
| TNK | Tyumenskaia Neftianaia Kompania (Tyumen Oil Company) | *Russia* |

## Abbreviations for Companies, Associations, & Organizations

| | | |
|---|---|---|
| TOM | Total Oil Marine plc | *UK* |
| TOP | Tedcastles Oil Products Ltd. | *Ireland* |
| TOR | Tema Oil Refinery | *Ghana* |
| TORC | Thai Oil Refinery Co. | *Thailand* |
| TOTAL | Total Fina Elf (merger of former CFP, PetroFina, Elf Aquitaine) | *France* |
| TPAO | Turkish Petroleum Corp. (Turkiye Petrolleri Anonim Ortakligi) | *Turkey* |
| TPDC | Tanzanian Petroleum Development Corp. | *Tanzania* |
| TPIC | Turkish Petroleum International Company Ltd. | *Turkey* |
| TPOC | Turkish Petroleum Overseas Company | *Turkey* |
| TRANSMED | Transmediterranean Pipeline Co. Ltd. (also TNPC) | *Algeria* |
| TRAPIL | Societe des Transports Petroliers Par Pipeline | *France* |
| TUMAS | Turkish Engineering Consultancy & Contracting Corp. | *Turkey* |
| TUPRAS | Turkiye Petrol Rafinerileri A.S. | *Turkey* |
| UCSIP | Union des Chambres Syndicales de l'Industrie du Petrole | *France* |
| UFIP | Union Francaise des Industries Petrolieres | *France* |
| UIE | Union Industrielle et d'Entreprise | *France* |
| UKOOA | UK Offshore Operators' Assoc. Ltd. | *UK* |
| UKPIA | UK Petroleum Industry Assoc. Ltd. | *UK* |
| URAG | Unterweser Reederei GmbH | *Germany* |
| VEBA | Veba Oel | *Germany* |
| VELA | Vela International Marine Ltd. | *Saudi Arabia* |
| VITOL | Vitol and global subsidiaries | *Netherlands* |
| VOST | Veba Oil Supply and Trading GmbH | *Germany* |
| WAPET | West Australian Petroleum Pty. Ltd. | *Australia* |
| WEC | World Energy Council | *UK* |

| | | |
|---|---|---|
| WEG | Wirtschaftsverband Erdol und Erdgasewinnung e.V. | *Germany* |
| WEPCO | Western Desert Petroleum Company | *Egypt* |
| WKTANRP | West Kazakhstan Territorial Administration for Natural Resource Protection | *Kazakhstan* |
| WOC | Wahe Oil Company | *Libya* |
| WPC | World Petroleum Congresses | *UK* |
| WTO | World Trade Organization | — |
| XPGC | Xiamen Petroleum Group Co. Ltd. | *China* |
| YANPET | Saudi Yanbu Petrochemical Co. | *Saudi Arabia* |
| YPC | Yemen Petroleum Company | *Yemen* |
| YPF | YPF S.A. (formerly Yacimientos Petroliferos Fiscales of Argentina) (see Repsol-YPF) | *Spain* |
| YPFB | Yacimientos Petroliferos Fiscales Bolivianos | *Bolivia* |
| YUKONG | S K Corporation | *Korea* |
| YUKOS | Derived from YUganskneftegas+ KuibishevnefteOrgSintez | *Russia* |
| ZADCO | Zakum Development Co. | *Abu Dhabi* |
| ZAO BNK | CJSC Baltiskaya Nefteprevalochnaya Kompaniya | *Russia* |
| ZRCC | Zhenhai Refining & Chemical Co. Ltd. | *China* |

# Miscellaneous Information and Symbols

## Common Oilfield Spellings

**A**
about–face
aboveground
acknowledgement
acre-feet
aeration
aftercooler
aftertreat
afterwash
air flow (n, adj)
airfoil
air line
airtight
alumna – s fem.
alumnae – pl gem.
alumnus – s masc.
alumni – pl masc.
anti-icer

**B**
Backflow
Backlight
Back off
Back pressure
Backup (adj)
Backwash
Backwater
Base line
Baseplate
Behavior
Belowground
Bench mark
Bench scale
Blow-by
Blowdown
Blowout
Boil off
Borehole
Bottom hole
Breakaway
Breakdown
Break-even (adj)
Break even (v)
Breakout
Breakpoint
Breakthrough
Breakup (n)
Break up (v)
Briquette
Bubble cap
Buildup (n)
Build-up (adj)
Build up (v)

Built in
Burn-off
Burnout (n)
Burnup (n)
Bypass
Byproduct

### C

Caribbean
carry-over
casinghead
center line
changeable
change out (v)
changeover
channeling
charge stock
checklist
check-out (adj)
check out (v)
checkpoint
city-wide
classroom
Cleanout
Cleanup (n)
Clean up (v)
Cleanup (adj)
Clear-cut
Close up (v)
Close-up (n, adj)
Closeout
Coastline
Commitment
commingle

controlling
controlled
co-op
co-owner
coproduct
counterbalance
counterbattery
countercurrent
counterflow
country-wide
crisscross
criterion – s
criteria – pl
cross- bedding
cross-country
cross flow
crosshead
crossover
cross plot
cross-reference
cross section
cutdown
cutoff
cutout
cut point
cycle oil

### D

datum – s
data – pl
deadman
dead time
deadweight
deepwater (adj)

## Miscellaneous Information and Symbols

deep water (n)
de-ethanizer
desiccant
desirable
dew point
doghouse
dogleg
double-jointed
downdip
downdrag
downflow
downgrade
downhole
downstream
downthrown
downtime
drawdown
drawoff (n)
drawworks
drill bit
drill collar
drillhead
drillpipe
drillship
drillsite
drillstem
drillstock
drillstring
drumhead

**E**
edgewise
end point
en route

**F**
face-lifting
fail-safe
falloff
farmin (n)
farmout (n)
feedback
feed rate
feedstock
feedwater
fiberglass
fieldman
fill-up
fireflood
firebox
fire wall
firewater (n)
flatbed
flier
flow-control
flow line
flowmeter
flywheel
foam glass
follow- up
forklift
formula
formulas
fourfold
freeze-up
freshwater (adj)
frost line
full time

## G
gamma ray
gas oil
gearbox
gearshift
grassroots
gray
ground line
groundwater
guesswork
guideline

## H
halfway
hammerblow
handwritten
hardback
headlong
hold-down
holdup
homemade
hookup

## I
in between (v)
in-between (n, adj)
industry- wide
infill
inflow (n)
in-line
inrush
iso-octane (no other "iso" words)

## J
jackknife
jackup (n)
jack up (v)
jobsite
judgment

## K
kerosene
know-how
knockout
knowledgeable

## L
landfill
landmass
lay barge
laydown (n)
lay-down (adj)
lay down (v)
leakoff
leakproof
left-hand
lengthways
letdown
lightweight
lineup (n)
line up (v)
linkup (n)
link up (v)
lowboy
lockout

## M
main line
mainstream

## Miscellaneous Information and Symbols

makeup
manageable
man-day
man-hour
man-year
mathematics
measurable
Mediterranean
Midcontinent
Mideast
midyear
mile-wide
millsite
multimillion-dollar
minable
modeling
mountainside
mousehole
movable
mud cake
mud line

### N

nationwide
nearby
non-communist
noticeable

### O

observable
oceanfront
offgas
off-line
off-loading
off site

offtake
oil field (n)
oilfield (adj)
oilman
oil well (n)
oil-well (adj)
onshore
on site (n)
on stream (n)
open Hole
outfall
overall
overhead
overpressure
overnight
override

### P

paperback
passthrough
payout
percent
phaseout
pickup
piggyback
pinchout
pipelay
pipelaying
pipelayer
pipeline
pipe rack
pipe still
powerhouse
predominant

## Standard Oil & Gas Abbreviator

printout
proof-test
pullout (n, adj)
pull out (v)
push button

### R

rainwater
rathole
readout
realizable
real time
reconnaissance
reentry
removable
right-of-way
rig-up (n)
rig up (v)
ringwall
riprap
roundoff
round trip
runback
rundown (n)
run down (v)
run-down (adj)
run forward
runoff

### S

salable
saltwater (adj)
salt water (n)
salvageable
sandblasting
Scale-up
Seabed
sealift
seawater
second hand (n)
secondhand (adj)
sendout
set point
setup (n)
severalfold
shipshape
shipside
shipyard
shoreline
shortcut
shortsighted
shortwave
shot hole
shot point
shut down (v)
shutdown (n, adj)
shut in (v)
shutin (n, adj)
shutoff (adj)
side boom
side cut
side draw
side stream
sidetrack
sidewall
sizable
slack line
slip joint
slipstream

## Miscellaneous Information and Symbols

slow-up
soleplate
spanwise
spectrum – s
spectra – pl
spot-check
standby
standoff
standpipe
standpoint
standpost
standstill
start-up (n, adj)
start up (v)
steamflood
steam line
step-out (n)
step-up (n, adj)
stepwise
straightforward
straight-run
stratum – s
strata – pl
stiffleg
stopgap
subpoena
superheat
switchgear

**T**

tailor-made
takeoff
takeover
take-up

tank car
tank truck
teardown
tie-in
tie-down
titleholder
toolpusher
toss-up
towline
trademark
trade name
trade off
trade out
transatlantic
traveling
trouble-free
troubleshoot
trunk line
turn down (v)
turndown (n, adj)
turnkey

**U**

ultraviolet
underwater
underway
updip
upflow
upstream
up-to-date
usable

**W**

warm-up (adj)
washout

waste water
water-cooled
water cut
waterflood
waterhead
waterline
waterside
watertight
waveform
wave front
wavelength
weathertight
wellhead
wellbore
wellsite
wellstream
wet out

whipstock
windblow
wind-chill (adj)
wireline
workboat
workbox
work load
workover (n, adj)
work over (v)
worldwide
wraparound

X-ray

year-end

Miscellaneous Information and Symbols

## API Standard
## Oil-Mapping Symbols

○ Location

erase symbol  Abandoned location

⌀ Dry Hole

● Oilwell

● Abandoned oil well

⌽ Distillate well

⌽ Abandoned distillate well

⦿ Dual completion—oil

◎ Dual completion—gas

∅w Drilled water-input well

●w Converted water-input well

∅G Drilled gas-input well

●G Converted gas—input well

○⋯× Bottom-hole location
    (x indicates bottom of hole. Changes in well
    status should be indicated as in symbol above)

⊕ Salt water disposal well

Courtesy American Petroleum Institute, Division of Production

## Mathematical Symbols and Signs

| | | | |
|---|---|---|---|
| + | plus | ∴ | there |
| − | minus | ∵ | because |
| ± | plus or minus | : | is to, divided by |
| × | multiplied by | :: | as; equals |
| . | multiplied by | ∺ | geometrical proportion |
| ÷ | divided by | ∝ | varies as |
| / | divided by | ≐ | approaches a limit |
| = | equal to | ∞ | infinity |
| ≠ | not equal to | ∫ | integral |
| ≈ | nearly equal to | $d$ | differential |
| ≅ | congruent to | ∂ | partial differential |
| ≡ | identical with | Σ | summation of |
| ≢ | not identical with | ! | factorial product |
| ≎ | equivalent to | π | pi (3.1416) |
| > | greater than | $e$ | epsilon (2.7183) |
| ≯ | not greater than | ° | degree |
| < | less than | ′ | minute; prime |
| ≮ | not less than | ″ | second |
| ≥ | greater than or equal to | ∠ | angle |
| ≤ | less than or equal | ∟ | right angle |
| ∼ | difference between | ⊥ | perpendicular |
| ≂ | difference between | ○ | circle |
| -: | difference between | ⌢ | arc |
| √ | square root | △ | triangle |
| ∛ | cube root | □ | square |
| ⁿ√ | nth root | ▭ | rectangle |

Miscellaneous Information and Symbols

## Greek Alphabet

| | | | | | |
|---|---|---|---|---|---|
| A | α | Alpha | N | ν | Nu |
| B | β | Beta | Ξ | ξ | Xi |
| Γ | γ | Gamma | O | o | Omicron |
| Δ | δ | Delta | Π | π | Pi |
| E | ε | Epsilon | P | ρ | Rho |
| Z | ζ | Zeta | Σ | σ | Sigma |
| H | η | Eta | T | τ | Tau |
| Θ | θ | Theta | Υ | υ | Upsilon |
| I | ι | Iota | Φ | φ | Phi |
| K | κ | Kappa | X | χ | Chi |
| Λ | λ | Lambda | Ψ | ψ | Psi |
| M | μ | Mu | Ω | ω | Omega |

## Frequently Cited Chemical Abbreviations

| | |
|---|---|
| CFCs | Chlorofluorocarbons |
| CO | Carbon Monoxide |
| NOx | A mixture of nitrous oxide and nitrous dioxide |
| PCBs | Polychlorinated biphenyls |
| TCDD | tetrachlorobenzo-p-dioxin |

Miscellaneous Information and Symbols

## Frequently Cited Additive Abbreviations

| | |
|---|---|
| ACHL | Alcohol |
| ACID | Acid |
| ACIN | Acid Inhibitors |
| ACRT | Acid Retarders |
| BIOC | Biocides |
| BRKR | Breakers |
| CACL | Calcium Chloride |
| CLYS | Clay Stabilizers |
| CO2 | Carbon Dioxide |
| CRIN | Corrosion Inhibitors |
| DFOM | Defoamers |
| DISP | Dispersion |
| DIVA | Diverting Agent |
| EMUL | Emulsifiers |
| FECL | Iron Control |
| FINS | Fines Suspender |
| FLA | Fluid Loss Agent |
| FRDC | Friction Reducers |
| GELA | Gelling Agents |
| GELO | Oil Gelling Agent |
| GLST | Gel Stabilizer |
| HCL | Hydrochloric Acid |
| HF | Hydrofluoric Acid |
| MAPE | Magnesium Pellets |
| MOTH | Moth Ball |
| MSLV | Mutual Solvents |
| NEML | Nonemulsifier |
| NTGN | Nitrogen |
| OSCV | Oxygen Scavengers |

| | |
|---|---|
| PFCL | Parrafin Control |
| PHBF | PH Buffers |
| PHOS | Phosphate |
| SALT | Salt |
| SCIN | Scale Inhibitors |
| SFAC | Surfactants |
| TROL | Tretolite |
| U | Unknown |
| WGEL | Water Based Gel Systems |
| WPLY | Water Based Polymers |
| XLNK | Cross-Linkers |

Miscellaneous Information and Symbols

## Frequently Cited Fluids Used as Cushion Abbreviations

| | |
|---|---|
| AMMON | Ammonia |
| $CO_2$ | Carbon Dioxide |
| MISRUN | Misrun |
| MUD | Mud |
| NTGN | Nitrogen |
| OIL | Oil |
| OILWTR | Oil and Water |
| PIPFLR | Pipe Failure |
| PKRFLR | Packer Failure |
| REVOUT | Reverse Out |
| TSTPLG | Tester Plugged |
| U | Unknown |
| WTR | Water |
| WTRNGN | Water and Nitrogen |

## Directional Survey Calculation Methods

- AA — Angle Averaging
- BT — Balance Tangential
- CA — Circular Arc
- ER — Exact Radius of Curvature
- MC — Minimum Curvature
- RC — Radius of Curvature
- T — Tangential
- U — Unknown
- VA — Vector Averaging

## Directional Survey Processing Types

I  Interpolated
M  Mixed
N  NonInterpolated

## Directional Survey Type or Method to Determine Wellbore Path Deviation

| | |
|---|---|
| AB | Acid Bottle |
| BGT | Borehole Geometry Tool |
| COMB | Combination Tool (Unspecified) |
| D | Dipmeter |
| EMS | Electronic Magnetic Multi-Shot |
| FMS | Formation Microscanner |
| G | Gyroscopic |
| GCT | Guidance Contiuous Tool |
| GMS | Gyroscopic Multi-Shot |
| MSS | Magnetic Single Shot |
| MWD | Measured While Drilling |
| SHDT | Stratagraphic High Resolution Dip Tool |
| SS | Single-Shot |
| TO | Totco |
| U | Unknown |
| WST | Wireline Steering Tool |
| GPIT | General Purpose Inclinometer Tool |
| GSS | Gyroscopic Single-Shot |
| GYRF | Finder Gyroscopic |
| GYRP | Pygmy Gyroscopic |
| GYRR | Unspecified Rate Gyroscopic |
| HDTD | High Resolution Dip Tool |
| HDTG | High Resolution Dip Tool |
| HDTS | High Resolution Dip Tool |
| HRD | High Resolution Dip Tool |
| M | Magnetic |
| MMS | Magnetic Multi-Shot |
| MS | Multishot |

Miscellaneous Information and Symbols

## Lithology Abbreviations

| | | | |
|---|---|---|---|
| AN | Anhydrite | LS | Limestone |
| AS | Ash | MR | Marlstone |
| CG | Conglomerate | MV | Massive |
| CH | Chat | NV | Novaculite |
| CK | Chalk | OL | Oolite |
| CO | Coal | OT | Other |
| CT | Chert | PR | Porosity |
| CY | Clay | QT | Quartzite |
| DN | Dense | RW | Reworked |
| DO | Dolomite | SH | Shale |
| DT | Detrital | SL | Slate |
| ER | Eroded | SO | Solid |
| GR | Granite | SS | Sandstone |
| GW | Granite Wash | UC | Silt |
| GY | Gypsum | WT | Weathered |
| HA | Salt | ZN | Zone |
| LG | Lignite | | |

## System Equivalents

### Metric Systems
**Length**   1 m = 100 cm = 1,000 mm = 0.001 km

**Area**   1 m$^2$ = 10,000 cm$^2$

**Volume**   1 m$^3$ = 1,000,000 cm$^3$ = 1,000,000 mL = 1,000 L

**Mass**   1 kg = 1,000 g
1 kg-mole = 1,000 gm-moles

**Density**   1 kg/m$^3$ = 0.001 g/cm$^3$

### English System
**Length**   1 ft = 12 in. = 0.333 yd = 0.000189 miles
1 mile$^2$ = 5,280 ft = 1,750 yd

**Area**   1 ft$^2$ = 144 in.$^2$
1 mile$^2$ = 27,878,400 ft$^2$

**Volume**   1 ft$^3$ = 1728 in.$^3$ = 0.178 bbl = 0.48 U.S. gal = 6.23 Imp. gal
1 bbl = 5.61 ft$^3$ = 42 U.S. gal = 34.97 Imp. gal

**Density**   1 lb/gal = 7.48 lb/ft$^3$ = 42 lb/bbl

Miscellaneous Information and Symbols

# Metric-English Systems Conversion Factors
## Basic Dimensions

### Metric System

**Length**  meter (m)
kilometer (km)
centimeter (cm)
millimeter (mm)

**Area**  square meters ($m^2$)
square centimeters ($cm^2$)

**Volume**  cubic meters ($m^3$)
cubic centimeters ($cm^3$)
liters (l)
milliliters (ml)

**Mass**  kilograms (kg)
grams (g)
gram-moles (gm-moles)
kilograms-moles (kg-moles)

**Density**  $kg/m^3$, $g/cm^3$

### English System

**Length**  inch (in.)
foot (ft)
yard (yd)
mile (mile)

**Area**  square inches ($in.^2$)
square feet ($ft^2$)
square miles ($miles^2$)

**Volume**  cubic inches (in.$^3$)
cubic feet (ft$^3$)
barrels (bbl)
U.S. gallons (gal)
Imperial gallons (Imp. gal)

**Mass**  pounds (lb)
pound-moles (lb-moles)

**Density**  pounds per gallon (lb/gal, lb/ft$^3$)

Miscellaneous Information and Symbols

## Basic Conversion Factors

**Length**
1 m = 3.281 ft = 39.37 in.
1 ft = 0.305 m = 30.5 cm = 3,050 mm
1 mile = 1.61 km
1 km = 0.621 mile

**Area**
1 $m^2$ = 10.76 $ft^2$ = 1,549 $in.^2$
1 $ft^2$ = 0.0929 $m^2$ = 929.4 $cm^2$

**Volume**
1 $m^3$ = 35.32 ft2 = 6.29 bbl
1 L = 0.035 $ft^3$ = 61 $in.^3$
1 $ft^3$ = 0.0283 $m^3$ = 28.31
1 bbl = 0.159 $m^3$ = 1591

**Mass**
1 kg = 2.205 lb
1 lb = 0.454kg = 454g
1 metric ton = 1,000 kg = 2,205 lb

**Density**
1 $kg/m^3$ = 0.0624 lb/ $ft^3$
1 $lb/ft^3$ = 16.02 kg/ $m^3$ = 0.01602 $g/cm^3$
1 $g/cm^3$ = 62.4 $lb/ft^3$

**Force**
1 kg force = 2.205 lb force
1 lb force = 0.454 kg force

**Work & Heat**
1 Btu = 0.252 kilocalories (kcal)
1 kcal = 3.97 Btu

**Power**
1 kilowatt (kw) = 860 kcal/hr = 3,415 Btu/hr = 1.341 horsepower (hp)
1 hp = 0.746 kw = 641 kcal/hr = 2,545 Btu/hr

**Enthalpy**
1 kcal/kg = 1.8 Btu/lb
1 Btu/lb = 0.556 kcal/kg

**Pressure**

1 bar = 14.51 lb/in.$^2$ (psi) = 0.987 atmospheres (atm) = 1.02 kg/cm$^2$
1 kg/cm2 = 14.22 psi = 0.968 atm
1 psi = 0.0703 kg/cm$^2$

**Temperature**

°C = 0.556 (°F − 32)
°K = °C + 273
°F = 1.8°C + 32
°R = °F + 460

Miscellaneous Information and Symbols

# Minerals Management Services (MMS) Two-Digit Area Prefix Standard

| | | | |
|---|---|---|---|
| AB | NG 15-8 | HE | Henderson |
| AC | Alaminos Canyon | HH | Howell Hook |
| AP | Apalachicola | HI | High Island Area |
| AT | Atwater | KC | Keathley Canyon |
| BA | Brazos Area | KW | Key West |
| BM | Bay Marchand Area | LL | Lloyd |
| BS | Breton Sound Area | LP | Lighthouse Point (field) |
| CA | Chandeleur Area | LU | Lund |
| CC | Corpus Christi | MA | Miami |
| CD | NG 16-7 | MC | Mississippi Canyon |
| CH | Charlotte Harbor | MI | Matagorda Island Area |
| CP | Coon Point | MO | Mobile |
| CS | Chandeleur Sound | MP | Main Pass Area |
| DC | Desoto Canyon | MQ | Marqesas |
| DD | Dustin Dome | MU | Mustang Island Area |
| DT | Day Tortugas | PA | NG 15-9 |
| EB | East Breaks | PB | St. Petersburg |
| EC | East Cameron Area | PE | Pensacola |
| EF | NG 16-8 | PI | Port Isabel |
| EI | Eugene Island Area | PL | South Pelto Area |
| EL | The Elbow | PN | North Padre Island Area |
| EW | Ewing Bank | PR | Pully Ridge |
| FM | Florida Middle Ground | PS | South Padre Island |
| GA | Galveston Area | RK | Rankin |
| GB | Garden Banks | SA | Sabine Pass (Louisiana) |
| GC | Green Canyon | SM | South Marsh Island Area |
| GH | NG 17-12 | SP | South Pass Area |
| GI | Grand Isle Area | SS | Ship Shoal Area |
| GV | Gainesville | ST | South Timbalier Area |

SX  Sabine Pass (Louisiana)
TP  Tarpon Springs
TS  Tiger Shoal
VK  Viosca Knoll
VN  Vernon
VR  Vermilion Area
WC  West Cameron Area
WD  West Delta Area
WP  NG 16-11
WR  Walker Ridge

Miscellaneous Information and Symbols

## **Petrophysical Curve Mneumonics**

The appropriate companies and associations have not established standard abbreviations for the petrophysical segment of the oil and gas industry. The following are for your convenience.

### **Wireline and LWD Curves**

| | |
|---|---|
| SP | Spontaneous Potential, mV |
| SPe | Spontaneous Potential edited (spliced, gap filled, environmentally corrected), GAPI |
| SPN | Spontaneous Potential normalized, mV |
| GR | Gamma Ray, GAPI |
| GRe | Gamma Ray edited (spliced, gap filled, environmentally corrected), GAPI |
| GRN | Gamma Ray normalized, GAPI |
| RHOB | RHOB, g/cc |
| RHOBe | RHOB edited, g/cc |
| NPHI | NPHI, decimal |
| NPHIe | NPHI edited, decimal |
| DRES | Deep Resistivity (spliced, gap filled), ohmm |
| DRESH | Deep Resistivity horizontal, ohmm |
| MRES | Medium Resistivity, ohmm |
| MRESH | Medium Resistivity horizontal, ohmm |
| SRES | Shallow Resistivity, ohmm |
| SRESH | Shallow Resistivity horizontal, ohmm |
| DT | Compressional slowness, us/ft |
| DT_v | Edited vertical compressional slowness, us/ft |
| Vp | Compressional velocity, km/s |
| Vp_v | Vertical compressional Vp, km/s |
| DTS | Shear slowness, us/ft |
| DTS_v | Edited vertical shear slowness, us/ft |
| Vs_v | Vertical shear Vs, km/s |
| CALI | Caliper, inch |
| CALI1 | Four-arm caliper, inch |

| | |
|---|---|
| CALI2 | Four-arm caliper, inch |
| PHITNMR | Total porosity from NMR log, decimal |
| PHBWNMR | Bound water porosity from NMR log, decimal |
| KNMR | Air permeability from NMR log, md |
| RFTPPG | Pore pressure gradient from RFT-type tool (MDT, RCI) lb/g |
| LOT | Leak-off test result, lb/g |
| MW | Mud weight, lb/g |
| BS | Borehole size, inch |
| PHIT | Total Porosity, decimal |
| PHIE | Effective Porosity, decimal |
| Sw | Water Saturation, decimal |
| PERM | Air Permeability, md |
| Ko | Oil (Gas) Permeability, md |
| Vcl | Clay volume fraction, decimal |
| Vsh | Shale (laminar) volume fraction, decimal |
| LITH | Lithology code (1, 2, 3, . . .) |
| PAY | Pay Flag |
| RES | Reservoir Flag |
| Vsps | Computed vertical shear Vs, km/s |
| AI | Acoustic Impedance, g/cc |
| SI | Shear Impedance, g/cc*km/s |
| VpVs | Vp/Vs ratio |
| PPG | Pore pressure gradient, lb/g |
| FG | Fracture gradient, lb/g |

### CORE and SWC Data

| | |
|---|---|
| COREPHIT | Core porosity, decimal |
| COREPERM | Core air permeability, md |
| COREVCL | Core total clay volume fraction (thin section or XRD), decimal |
| COREDM | Core mean grain size from LPSA, um |
| SWCPHIT | SWC porosity, decimal |
| SWCDM | SWC mean grain size from LPSA, um |
| SWCPERM | Computed SWC air permeability, md |

## PennWell Software License Agreement

THIS IS A LEGAL AGREEMENT BETWEEN YOU AND PENNWELL. CAREFULLY READ ALL THE TERMS AND CONDITIONS OF THIS AGREEMENT. INSTALLING THIS CD-ROM INDICATES YOUR ACCEPTANCE OF THESE TERMS AND CONDITIONS.

1. License: This License Agreement grants the purchaser of this package one License to use one copy of the specified version of the enclosed PennWell product ("Software") on any single computer. You may transfer the Software from one computer to another so long as it is not used on more than one computer at a time. YOU MAY NOT TRANSMIT THE SOFTWARE OR THE DATA CONTAINED WITHIN FROM ONE COMPUTER TO ANOTHER IN A NETWORK OR TO SERVE MULTIPLE USERS WITHOUT PURCHASING A SITE LICENSE FROM PENNWELL. Solely for your own backup purposes, you may make a single copy of the Software in the same form as provided to you on the enclosed CD.

2. Transfer: You may not transfer the License to another party. As a buyer of this Software you agree that the DATA contained on the CD will be used by employees of and within the corporate structure of the BUYER and will not be disclosed to any organization or person, subsidiary or affiliate of the Buyer. You have no right to sublicense or loan the Software.

3. Copyright: The software and the Installation & Configuration are owned by PennWell. PennWell holds all right, title, and interest in and to this product. The Software and Documentation are protected by United States and international copyright laws and international treaty provisions. You may not remove, obscure, or alter any notice of patent, copyright, trademarks, trade secret, or other proprietary rights.

4. Terms: This License is effective until terminated. This License and your right to use the Software terminate automatically if you violate any part of this Agreement. You agree upon termination to return or destroy within 5 days all copies of the Software and to affirm in writing to PennWell that you have done so.

5. Limited Warranty (Disclaimer and Limitation of Liability): PennWell warrants the enclosed CD on which the Software is provided to be free from defects in materials and workmanship at the time of delivery to you.

PennWell has made reasonable checks of the Software to confirm that it will perform in normal use on compatible equipment substantially as described in PennWell specifications for the Software, as published most recently prior to the delivery of this package to you. However, due to the inherently complex nature of computer software, PennWell does not warrant that the Software or the Documentation is completely error-free, will operate without interruption, is compatible with all equipment and software configurations, or will otherwise meet your needs. ACCORDINGLY, THE SOFTWARE AND DOCUMENTATION ARE PROVIDED "AS-IS," AND YOU ASSUME ALL RISKS ASSOCIATED WITH THEIR USE.

AS YOUR SOLE REMEDY FOR ANY BREACH OF WARRANTY, you may return to PennWell the original copies of the Software and Documentation, along with proof of purchase and any backup copies, for replacement or (at PennWell choice) for a refund of the amount you paid for this package, provided the return is completed within 90 days following the delivery of this package to you.

PENNWELL MAKES NO OTHER WARRANTIES EXPRESSED OR IMPLIED, WITH RESPECT TO THE SOFTWARE OR THE DOCUMENTATION, THEIR MERCHANTABILITY, OR THEIR FITNESS FOR ANY PARTICULAR PURPOSE. ALL WARRANTIES, EXPRESS OR IMPLIED, WILL TERMINATE UPON THE EXPIRATION OF 90 DAYS FOLLOWING DELIVERY OF THIS PACKAGE TO YOU. Some states do not allow limitations on how long an implied warranty lasts, so the above limitation may not apply to you.

IN NO EVENT WILL PENNWELL BE LIABLE FOR INDIRECT, INCIDENTAL, OR CONSEQUENTIAL DAMAGES, INCLUDING WITHOUT LIMITATIONS, LOSS OF INCOME, USE, OR INFORMATION, NOR SHALL THE LIABILITY OF PENNWELL EXCEED THE AMOUNT PAID FOR THIS PACKAGE. Some states do not allow the exclusion or limitation of incidental or consequential damages, so the above limitation or exclusion may not apply to you. This warranty gives you specific legal rights, and you may also have others, which vary from state to state.

6. General: This Agreement constitutes the entire Agreement between you ("the Buyer") and PennWell and includes the Terms and Conditions as outlined in this agreement.

   Technical Support: PennWell agrees to offer limited support with the initial installation and the use of the product should help be necessary.

   The Software does have a built-in help system that should answer most questions. Phone support will be available and restricted to the Buyer of the Software as outlined in this agreement.

7. Export: Distribution of the PennWell Software is subject to compliance with all laws, regulations, orders, and other restrictions on export from the United States of America of the PennWell Software or any technical information about the PennWell Software that are imposed by the government of the United States of America.

8. U.S. Government Restricted Rights: The Software and Documentation are provided with RESTRICTED RIGHTS. Use, duplication, or disclosure by the Government is subject to restrictions as set forth in subparagraph (c)(1)(ii) of the Rights in Technical Data and Computer Software clause at DFARS 225.2270-7013 or subparagraphs (c) (1) and (2) of the Commercial Computer Software Restricted Rights at 48 CFR 542.227-19, as applicable.

9. PennWell Address: Contractor/manufacturer is
   PennWell, 1421 S. Sheridan Road, Tulsa, OK 74112, (918) 835-3161

## System Requirements

### Windows Version
- Intel® Pentium® processor
- Microsoft® Windows® 95, Windows®98, Windows NT®* 4.0 with Service Pack 5 or 6, or Windows 2000 (*Internet Explorer 4.0.1 or later required for Windows NT users)
- 32 MB of RAM (64 MB recommended)
- 115 MB of available hard-disk space
- CD-ROM drive

### Macintosh Version
- PowerPC® processor
- Mac OS software version 8.6*, 9.0.4, or 9.1 (*Some features may not be available due to OS limitations)
- 32 MB of RAM (with virtual memory on) (64 MB recommended)
- 105 MB of available hard-disk space
- CD-ROM drive

## Installation

In order to use this CD ROM you must first have Acrobat® Reader 5.0 installed on your computer. If not, then use the following instructions to run the Acrobat Reader 5.0 auto-install program that is best suited for your operating system.

### Windows
- Insert the CD into the CD-ROM drive.
- Click "Start" on the task bar and the select "Run."
- In the "Run" dialogue box, choose "Browse" and from there open the CD from your CD-ROM drive.
- Double click the Installs folder. Double click on appropriate Install for your operating system. Click "OK" to begin installation.
- Follow the instructions in the Acrobat Reader 5.0 set-up dialogue box. Click "OK." This installs the necessary files to operate your CD.
- Next, locate the "DD5ED.pdf" file on the CD and double click to begin using this CD

### Macintosh
- Insert the CD into the CD-ROM drive.
- Open the CD and double click on the "Install Me First" file. This will cause the auto-install program to begin the installation process. Simply follow the instructions in Reader installer to install the program on your computer.
- Next locate the "DD5ED.pdf" file on the CD and double clock to begin using the CD.